MW01050168

Rare and Endangered Plants of Oregon

DONALD C. EASTMAN

Beautiful America Publishing Company

This book is dedicated to the memory of

SIBYL

She loved to search
for the very rare ones

especially the Fritillaria

Beautiful America Publishing Company©
P.O. Box 646
Wilsonville, Oregon 97070
(503) 682-0173

Printed in Hong Kong

Library of Congress Catalog Card Number 90-149

ISBN 0-89802-524-9
ISBN 0-89802-561-3 Hardbound

CONTENTS

VENGARICK
9731 SE LINWOOD AVE
MILWAUKIE, OR 97222
503-775-6563

Foreword

Don Eastman has explored many miles of back roads, waded into muddy marsh lands, and climbed the steep slopes of Oregon's mountains to take these photographs of our rare, threatened and endangered plants.

A native of Nampa, Idaho, Don practiced dentistry in McMinnville, Oregon, throughout most of his professional life. But he was also an avid hiker and mountain climber, and on his many trips into the back country, he was fascinated with the wildflowers he saw along the way.

Although enamored of mountains since boyhood, Don did not make his first climb until 1954, when he and Jim Craig, a lawyer from McMinnville, climbed Mt. Hood. They have since climbed most of the peaks in the Northwest and many in Canada and Europe.

I first met Don in 1969 when he received the first 50-peak award ever given by The Mazamas, the climbing club based in Portland, Oregon. I was impressed with his dedication and ability. In 1974, as Chairman of The Mazamas Climbing Committee, he signed my "Sixteen Major Northwest Peaks" award for climbing the 16 major peaks of the Cascade and Olympic Mountains from Mt. Shuksan in Washington to Mt. Shasta in California, including Mt. Rainier. Mine was only the 216th plaque given in the 80 year history of the Club, but by then Don had not only climbed all 16, but also had led climbs on each one of them! In 1970 he received the 100-peak award, the first and only one ever given by The Mazamas.

Throughout his many hikes and climbs, Don became increasingly interested in wildflowers. He began to photograph them to record their beauty, and found a whole new world through the close-up lens. But he was having so much difficulty finding someone who could identify the plants that he vowed someday to publish a book of photographs so others would know the wildflowers they were seeing. He has kept his vow. Early in his climbing career, it was the climbs that determined what plants he would see. Now it is the plants that determine where he will climb.

After he retired in 1981, he set out to photograph as many of the 4000 plants in Oregon as possible. He has combed the state every year since, driving as many as 25,000 miles a year and hiking many additional miles in search of them. He now has photographed over 2000 species. Those which are rare have been given top priority. Photographs of many of these have never before been published.

In 1982, he joined the Native Plant Society of Oregon, was the Vice-President, then President of the Willamette Valley Chapter in 1984 and 1985. He also served on the State Board of Directors for four years. He has given many talks, showing his slides, and has participated in wildflower shows.

That same year, Don became a volunteer with the Oregon Rare and Endangered Plant Project, of which I am the Director. Don has contributed greatly to this effort to determine which plants are rare, threatened and endangered in Oregon, sharing both information about where he had seen plants as well as photographs of the species.

This monumental book is a comprehensive document which brings together the information gathered during the past sixteen years by over 300 amateur and professional botanists in Oregon. Don has combined both the art of photography and the science of botany. He knows what features are needed for identification and has photographed them. This sets his pictures apart. His book will serve as a reference for both the serious botanist and the amateur. It is written in a language we all can understand.

Enjoy the volume, but know it takes thousands of miles and a lot of perseverance and dedication to produce such a book. To be an amateur means "to love" something and Don does, with heart, pen, and lens.

Jean L. Siddall
Lake Oswego, Oregon
August, 1989

A NEW AND RARE MARIPOSA-LILY

Crinite mariposa-lily
Calochortus coxii Godfrey & Callahan
Discovered in June 1988 by Marvin Cox on a serpentine slope between Myrtle Creek and Boomer Hill a few miles southwest of Roseburg. It has now been listed as Endangered throughout its limited range of about eleven square miles.

ACKNOWLEDGEMENTS

Through the many years it has taken to accumulate and put together the photographs and material for this book, the author has had the privilege of meeting and working with many very fine people. All of them, now considered friends, have been most willing to provide information and help in the search for many of the rarest, hardest to find plants in the state. Much appreciation is accorded the following people: **Jean Siddall**, state chairman of the Oregon Native Plant Society's committee for Rare and Endangered plant species of Oregon, and a co-author of Oregon's first published list of rare plants, "*Rare, Threatened and Endangered Vascular Plants in Oregon—An Interim Report*", published in October 1979, who freely gave access to her many files of siting records for rare wildflowers, for her help in identification, and especially for the time spent in editing, and in writing the Forward for this book. **Jimmy Kagan**, **Sue Vrilakas**, and **Cathy Macdonald** of The Nature Conservancy and Oregon Natural Heritage Data Base, publishers of "*Rare, Threatened and Endangered Plants and Animals of Oregon*" in 1983, 1985, 1987, and 1989, for information relating to recent sitings of rare plants, newly found species, changes in nomenclature, and help in identification. **George Lewis** of the Native Plant Society in Portland for sharing his invaluable and extensive knowledge of plant locations throughout the state. **Wilbur Bluhm**, **Clint Urey**, **Susan Kephart**, **George** and **Harriet Schoppert**, and **Larry Scoffield**, all from the Willamette Chapter of The Native Plant Society for help in locating and identifying many, many of the rare species of the Willamette Valley, Cascades, and Siskiyou Mountains. **Paula Brooks** and **Stephanie Schultz**, then of the Salem BLM office, for taking the author to a site for *Iris tenuis* and for instructions for finding *Corydalis aquae-gellidae*. **Keith Chamberlain** of Mosier for help with Columbia Gorge flowers, **Karl Urban** and **Bruce Barnes** of Pendleton for numerous plants of the Wallowas, Blue Mountains, Steens, and John Day country of central Oregon. **Lois Kemp** of Portland for her help in locating *Howellia aquatilis* in the slough area of the Columbia River near Ridgefield, Washington and for directions to the newly found *Sisyrinchium sarmentosum* in the Mt. Hood area. **Rachel** and **Roy Sines** of LaGrande for help with Wallowa species, including the endangered *Mirabilis macfarlanei* in the remote Imnaha River canyon. **Royal Swanson**, farmer near North Powder, for taking us to the rare *Thelypodium howellii var. spectabilis* which grew in a Nature Conservancy preserve on his ranch. **Cheryl McCaffrey**, then of the BLM office in Burns for help with *Stephanomeria malheurensis* and many species on Steens Mountain. **Stu Garrett**, **Dave Danley**, and **Marge Ettinger** in Bend for help in finding and identifying species of central Oregon. **Christy Steck** of Prineville for her guidance to many rare species in the Ochocos, Big Summit Prairie, and the Painted Hills. **Priscilla Mosser Eastman** for her help in finding rare plants in the central Oregon Cascades and in the Columbia Gorge. **Joan Seevers** of Medford BLM office, **Frank Lang** of Southern Oregon State College, **Lee Webb** of Siskiyou National Forest in Grants Pass, **Frank Callahan** of Gold Hill, all of whom went out of their way to show the author rare plants or give directions for finding them in the Siskiyous of southwest Oregon. **Bob Mansfield**, now deceased, for his help with rare *Fritillaria* and *Arabis* in the Siskiyous. **Larry Loftus** of Medford for guidance to the rare plants in the Miller Lake area. **Boyd Kline** of Medford for help with the many beautiful lilies of southwest Oregon. **Mr. and Mrs. Dale Hatch** of Foots Creek in Josephine County for leading us up a steep mountain to a most wonderful stand of *Cypripedium montanum* and *Cypripedium fasciculatum*. **Veva Stansell** of Pistol River, **Rosamund Hess** of Gold Beach, and **Helen Planeto** of Agness for their help in locating and identifying many rare plants along the south Oregon coast, the Rogue River, and the Siskiyou Mountains. **Alan Curtis**, **Rhoda Love**, **Charlene Simpson**, **Freeman Rowe**, **Wendell Wood**, all of Eugene, very helpful in finding rare species in that area. **Russ Holmes** and **Mildred Thiele** of Roseburg for help in finding rare plants of the North Umpqua. **Ramona Osborn** *of Jacksonville who led us to Fritillaria gentneri* growing in the woods back of her home. **Leo Torba** of Klamath Falls who gave directions for finding *Nama lobbii* and *Asarum wagneri* in Lake of the Woods area. **Morris Johnson** of Western Oregon State College in Monmouth for help in locating certain plants along the coast and in the Coast Range, and for help with the identification of others. **Bob Meinke**, new Program Coordinator for the Endangered Species Office, Plant Division, Department of Agriculture in Salem, for directions to certain rare species in the Wallowas and for help with identification. **Carolyn Wright** of Dufur for directions in locating the recently found *Eriogonum chrysops* far out in the wilds of Malheur County. In addition to these many fine people active in the field of botany, great appreciation is given to Beautiful America Publishing Company of Wilsonville, Oregon, to **Ted** and **Beverly Paul**, owners and managers of Beautiful America, for accepting this material for publication.

Introduction

Purpose of the Book:

Oregon has long been recognized for its scenic grandeur and great variety of geographical features, from the Pacific Ocean on the west, through the Coast Range and the Siskiyou Mountains to the great inland valleys of the Willamette, Umpqua, Rogue, and Illinois river drainages, up the magnificent Columbia Gorge, across the high backbone of the state, the Cascade Mountains, into eastern Oregon. There the diversity continues east through much drier country, across the Deschutes, John Day, Umatilla river drainages of the Columbia Plateau to a relatively high country to the south recognized as a part of the Great Basin which extends eastward and southward into Idaho, Utah, Nevada, and eastern California. Interesting and beautiful mountain ranges break the contour of eastern Oregon. The Ochoco Mountains in the center of the state, the Blues and the Wallowas in the northeast section, Steens Mountain, Pueblo and Trout Creek mountains in the southeast corner and the Warner and Gearhart ranges in south central Oregon. Cutting through these mountain ranges are tremendous gorges and canyons such as the Owyhee, the Imnaha, the Grande Ronde, and the deepest of all the canyons on the continent, Hells Canyon, cut by the mighty force of the fast-flowing Snake River along the eastern boundary of the state. This beauty, along with the wealth of recreational opportunities inherent in the state, has attracted people for over a hundred and fifty years. Many came to stay and live their lives here. They have come from all over America, from many parts of the world to see and feel and live the "Oregon Experience."

Inspite of this "experience" one of Oregon's greatest heritages has gone relatively unknown, the native plants that grow wild in this state. Because of the great geographical variety, there is a great natural diversity in the vegetation cover. In all the fifty states in America, Oregon ranks third in number of native species—well over four thousand. Only California and Texas, both much larger in area, have more. Each area of the state has its own distinct species, some, of course, spilling over into adjacent states. Most people are aware of many of the more prominent, obvious, and numerous plants such as our Douglas-fir, ponderosa pine, rhododendron, sage brush, and juniper, but few have seen the many unusual, rare plants of the state.

The author first became aware of a small part of this "treasure" as he hiked and climbed in the Cascade and Wallowa mountains. An early interest led to an attempt to count the many different flowers seen along a particular trail, without any idea as to what they were. Query to fellow climbers and hikers netted little or no gain in knowledge. Study of the plants through a camera lens intensified the interest, and macro-photography followed shortly. Eventually contact was made with organizations interested in native plants, such as The Native Plant Society of Oregon, the Oregon Rare and Endangered Plant Project, and The Nature Conservancy. Only then was progress made in the knowledge and understanding of some of these plants. The more one learns, the more one realizes the lack of knowledge, which all leads to a greater interest and effort in pursuing the subject.

All of this eventually brought about a disturbing discovery. Not just some, but many of the most attractive flowers in the state were diminishing in numbers. Some have even been declared extinct, or possibly so, within the state.

For example, the striking wild orchid known as the yellow lady's-slipper, *Cypripedium calceolus* L. *var. parviflorum* (Salisb.) Fern., has not been reported in Oregon since last seen in the Warner Mountains in the mid nineteen-thirties. One or two sites are still known in adjacent Washington. The author's photo of it (probably not the same variety) was taken above Lauterbrunnen, Switzerland, where it is also very rare and protected. The lovely small tiger lily of southwest Oregon, *Lilium parvum* Kell., appears to have disappeared from Oregon. Robinson's onion, *Allium robinsonii* Hend., once found on the banks immediately above the Columbia River is no longer there, a victim of the high water caused by the dams. It is now known only from the last remaining free-flowing stretch of the river near Hanford, Washington. The colorful Willamette daisy, *Erigeron decumbens* Nutt. *ssp. decumbens*, and the small but striking tri-colored monkey-flower, *Mimulus tricolor* Hartw. ex Lindl., were once common in the Willamette Valley but are now restricted to a few isolated sites. MacFarlane's four-o'clock, *Mirabilis macfarlanei* Const. & Roll., one of the rarest plants in America, is known only in a few extremely small populations in the Imnaha and Snake River canyons. Golden buckwheat, *Eriogonum chrysops* Rydb., a small, sensitive plant reported in the "vicinity" of Steens Mountain early this century, was miraculously rediscovered this year on a remote tableland far back in the wilds of Malheur County. This represents only a minute portion of the plants now endangered in Oregon. Rarities and extinctions of this type are often caused by human beings. Development of the land, logging of the forests, damming the rivers, picking, digging, and trampling of the more attractive species, all lead to their demise. Their destruction more often results from the lack of knowledge as to their presence, what they are, and sometimes an attitude of indifference—just something in the way of economic development.

The prime purpose of this publication is to acquaint Oregonians, and the visitors to Oregon, with a portion of this unique heritage of botanical treasure and the need to protect it. As people learn more, their interest is whetted and their desire to preserve what is left becomes keen. This in turn can lead to more effective protection and a preservation for future generations to also enjoy. Additionally, a plant lost to extinction is lost forever. It will never be known how useful it could have been to mankind. For example, over fifty percent of all pharmaceuticals are, or were, originally derived from plants and less than three percent of the world's plants have been chemically analyzed for use by mankind. It is essential to establish a compromise between conservation and development that will preserve these valuable entities before it is too late!

To date no photographic book has been published of Oregon's rare plants. Wildflower books limited to certain geographical portions of the state, and some regional publications have included pictures of some of Oregon's rare plants, but many of them have never before been published photographically. Several of the plants shown are recently discovered species or varieties.

The Plants Included:

Over the past eight years the author has made every effort to seek out and photograph the rarest plants in Oregon. Many of them are included in this book; many more remain out there to be found and incorporated into some subsequent edition. In this book are those species considered to be endangered in Oregon, or are now or have been on some rare plant list in this state, plus one or two plants listed as rare or endangered in adjacent states but not currently in Oregon. Some of the plants included have been on endangered, threatened, or watch lists in the past but because they are more abundant than previously thought, or because there is some uncertainty regarding nomenclature or identity, they have been dropped. The author does not necessarily concur with such decisions, as many of these plants are still rare. Plants that are dropped should continue to be on a "watch list" for a considerable length of time to assure they do not become threatened or endangered through carelessness and lack of knowledge of their rarity. Some lilies of southwest Oregon, such as Vollmer's and Bolander's, are examples. Both, though currently plentiful, are diminishing in numbers. They are extremely attractive to collectors who, knowing they are no longer listed as rare or endangered, do not hesitate to dig them at will. Flowers dug and transplanted usually do not last long in a different environment. Flowers picked can not produce seed, and may eventually be lost forever. For those who wish to have them in their gardens, seeds and bulbs can often be obtained from stock grown in nurseries. Those in the wild should never be touched. Take only photographs. Pictures will last much longer. Attractive print enlargements can be made and enjoyed for years, and will serve to remind one of the wonderful experiences of finding these beautiful creations in the wild. It may serve well to quote from an Oregon Native Plant Society brochure printed in 1988, "Of the approximately 20,000 species of native plants in the continental United States, at least ten percent are in danger of extinction, according to a study completed in 1978 by the Smithsonian Institution. In Oregon it is estimated that eleven plant species face extinction each year. Over thirty Oregon plants have not been seen since 1960 and are now considered extinct." It has been said, "Destroying a species before learning its worth is like burning a library before reading the books." To quote Dr. E. O. Wilson of Harvard University, "The one process ongoing in the late 1980's that will take millions of years to correct is the loss of genetic and species diversity by the destruction of natural habitats. This is the folly our descendants are least likely to forgive us;" and James L. Buckley, Undersecretary of State, puts it this way, "As living creatures, the more we understand of biological processes, the more wisely we will be able to manage ourselves. Thus the needless extermination of a single species can be an act of recklessness. By permitting high rates of extinction to continue, we are limiting the potential growth of biological knowledge. In essence the process is tantamount to bookburning; but it is even worse, in that it involves books yet to be deciphered and read." To repeat, a plant that becomes extinct is lost forever!

Conservation of Rare Plants:

Early in the 1970's, interest was focused by a few on the plants rarely seen in Oregon and those restricted to small geographical areas. Dr. Kenton L. Chambers, curator of the Oregon State University Herbarium, was the first to list these plants based on species rarely collected in Oregon. About the same time, Jean L. Siddall was preparing a list for the *Pacific Northwest Research Natural Area Committee*, based on what botantists considered to be rare and endangered in their area of the state. When Chambers and Siddall realized they were listing the same species using two different methods, they pooled their efforts. Impetus to this project was given by the passage in Congress of the Endangered Species Act in 1973 which included plants for the first time. With the publication in the the *Federal Register* of rare and endangered plant and animal species of the United States in July 1975 by the U. S. Fish and Wildlife Service, an Advisory Committee of professional taxonomists of the Oregon Rare and Endangered Plant Project was appointed to draw up Oregon's original list of rare, threatened, and endangered plants. It was published by the Oregon Natural Area Preserves Advisory Committee to the State Land Board in October 1979 as *Rare, Threatened and Endangered Vascular Plants in Oregon—An Interim Report*.

Early in this decade a computerized data bank on Oregon's rare plant and animal species was established by the *Oregon Natural Heritage Data Base* in the Oregon field office of The Nature Conservancy. In cooperation with The Native Plant Society of Oregon, The Oregon Department of Fish and Wildlife, The Oregon Natural Heritage Advisory Council, the state Rare and Endangered Plant Project, The United States Fish and Wildlife Service, and many of the leading botanists and zoologists of Oregon, updated listings were made and published as *Rare, Threatened and Endangered Plants and Animals of Oregon*, July 1983. Regularly scheduled conferences have followed, resulting in updates of this publication in March 1985 and April 1987. The latest revision was published in April 1989. (The changes from that revision are included in Appendix I). Specific sitings of many of Oregon's rare plants are on file and computer disk. This information is used for monitoring threatened and endangered species. The State Land Board, in March 1987, gave recognition to these lists as "identifying the status of rare, threatened and endangered plants and animals, and for use in facilitating research and monitoring efforts."

Unfortunately, these measures have still not provided any real protection for Oregon's truly rare plants. Because of that, the Native Plant Society of Oregon, through its Legislative Committee chaired by Esther McEvoy, took on the herculean task of getting legislation passed for this protection. This effort was successful in 1987 through the passage of Oregon Senate Bill 533. With its passage, responsibility for plants in Oregon which are endangered or threatened throughout their range is vested in the Oregon Department of Agriculture. Subsequently, botanist Bob Meinke of Oregon State University was appointed as Program Coordinator for The Oregon Endangered Plant Species Program in the Division of Natural Resources. This new division has been charged to compile a state list of threatened and endangered plant species, and to ultimately develop conservation and recovery plans for any listed taxa. The law provides for a preliminary review of appropriate candidate species in 1988. A species can be listed only if it is threatened or endangered throughout all of its geographic range. This necessarily includes only Federal Candidate species which tentatively could amount to a portion of the 170 to 180 taxa found on List 1 of the Oregon Natural Heritage Data Base list of 1987, plus a selected few not presently on the Federal list such as recently discovered species. Field surveys are being conducted by qualified botanists. Conservation programs will be developed for those plants listed. Close cooperation with the Forest Service, the Bureau of Land Management, and the Fish and Wildlife Service will be maintained through the provisions of the Federal Endangered Species Act. Annual workshops, newsletters, and contact and cooperation with adjacent state agencies will be a part of this program.

Adjacent state agencies involved with similar listings are the Washington Natural Heritage Program in Olympia, Washington, the Rare and Endangered Plants Technical Committee of the Idaho Natural Areas Council in Moscow, Idaho, the Nevada Natural Heritage Program in Carson City in cooperation with the Northern Nevada Native Plant Society, and the California Native Plant Society in Sacramento, California.

To date only three Oregon species have been listed as Endangered on the Federal Register by the Fish and Wildlife Service: Macfarlane's four-o'clock, *Mirabilis macfarlanei*, and Malheur wire-lettuce, *Stephanomeria malheurensis*, and just this year, Bradshaw's desert parsley, *Lomatium bradshawii*. A fourth species, Red Mountain rock cress, *Arabis macdonaldiana*, is on the Federal Register as being Endangered in California. Some claim that this species is also found in Curry County, Oregon; others say the Oregon plant is not the same, possibly a new species. Another seven plants are Category One species for Endangered status, meaning that "sufficient information is on hand to support the biological appropriateness for their being listed as Endangered or Threatened species." These are: *Hackelia cronquistii, Hastingsia bracteosa, Limnanthes floccosa ssp. pumila, Lomatium greenmanii, Luina serpentina, Senecio ertterae,* and *Thelypodium howellii ssp. spectabilis*. Lack of data "concerning environmental and economic impacts on listings" has held up action on these important rare plants for over ten years.

Another 135 Oregon species are listed as Category Two candidates for Endangered status, for which insufficient information is presently available to biologically support their listing. Here again, too much time is being taken to acquire this information. Hopefully, the new Oregon legislation can correct some of the inadequacies of the Federal Endangered Species Act in this state.

The new State Coordinator, Meinke, has already come up with a list of nineteen endangered or threatened Oregon species to be proposed as the first plants for listing by the State Department of Agriculture. They are: *Abronia umbellata ssp. breviflora, Amsinckia carinata, Astragalus applegatei, Calochortus umpquaensis, Erigeron decumbens, Haplopappus radiatus, Lilium occidentale, Lomatium bradshawii, Lomatium cookii, Lomatium greenmanii, Luina serpentina, Mentzelia packardiae, Mirabilis macfarlanei, Plagiobothrys hirtus, Pleuropogon oregonus, Senecio ertterae, Sidalcea nelsoniana, Stephanomeria malheurensis,* and *Thelypodium howellii ssp. spectabilis*. Meinke's philosophy for program effectiveness has been molded through several years of work with federal agencies. It may be summed-up by the following quotation: "Many people feel that we should list considerably more species than we are presently proposing, but we need to remember that the law requires us to demonstrate sound biological and ecological reasons for designations of threatened or endangered. The species we have selected for our intial list are either those for which we have convincing data, or (in a few cases) some which have such clearly compromised habitats, coupled with reduced population numbers, that we feel they qualify. We can accomplish our goals of conservation and species preservation much more effectively if we don't weigh our lists down with questionable species that may very possibly prove, in time, to not be threatened at all. . . this often serves to alienate the very land managers whose cooperation and support we need if our efforts are to be successful. Listing determinations have to be based on valid supporting data or we end up trivializing our methodology, thereby leaving ourselves open to charges of subjectivity. These decisions must be able to stand up to close professional scrutiny, even in court if necessary, or we may risk the program's effectiveness. It is, however, very important to keep track of all potentially sensitive species in some fashion. . . this is where environmental groups such as the Oregon Native Plant Society become so important."

Distribution of Rare Plants in Oregon:

Rare and endangered plant species occur in all geographical regions, and in all plant habitats throughout Oregon. Locations for plants included in this book are given in general terms only, as it is not meant to be a guide to specific sites. The small locater maps that accompany each plant show generally where a plant may be found within the state. Some plants cover large areas, but are very spotty in their occurrence within that large area. Most of the rare plants, however, occur in a relatively few sites that are small in area, and small in plant population. There are exceptions. For example, *Erythronium elegans* is found in very few sites, but in one of those sites it is estimated to number over 500,000. It is still considered to be a vulnerable species because of the small number of sites.

The *Interim Report* of Siddall, Chambers, and Wagner, 1979, lists 19 regions of botanical interest in the state. Listed according to the seven geographical sections of the state they are:

1. **Northwestern Oregon**: North Coast, the Coast Range, Willamette Valley, the North Cascades.
2. **Southwestern Oregon**: South Coast, Umpqua Valley, Siskiyou Mountains including the Rogue and Illinois river valleys, the South Cascades.
3. **North Central Oregon**: Columbia River Gorge and Columbia Basin.
4. **Central Oregon**: Ochoco Mountains, John Day Valley.
5. **South Central Oregon**: Klamath Basin, the Basin and Range areas of Lake and Harney counties.
6. **Northeastern Oregon**: The Blue Mountains, the Wallowa mountains, and Snake River Canyon.
7. **Southeastern Oregon**: Steens Mountain, the Owyhee Uplands.

The state has been divided also into seven basic vegetation types, after Detling, in 1968:

1. **Moist conifer forest** with abundant rainfall, up to 140 inches per year, and mild winter and summer temperatures.
2. **Boreal forest** with 25-60 inches of rain/snow per year and cold temperatures in the winter, milder along the coast.
3. **Alpine fell-fields** above timberline in the mountains with low winter and summer temperatures.
4. **Pine-oak forest**, dry (15-30 inches of rain per year), cold in winter and warm in summer.
5. **Chaparral**, dry (12-20 inches of rain), very dry in summer, with mild winters and hot summers.
6. **Juniper-sagebrush woodland**, very dry (7-13 inches of rain per year) with cold winters and warm summers.
7. **Grassland**, dry (7-18 inches of rain per year, continuing into late spring. The winters are cold and the summers warm.

Both the geographic regions of botanical interest, and the areas of major vegetation types in Oregon are shown on the accompanying map. See figure 1.

Regions of Botanical Interest

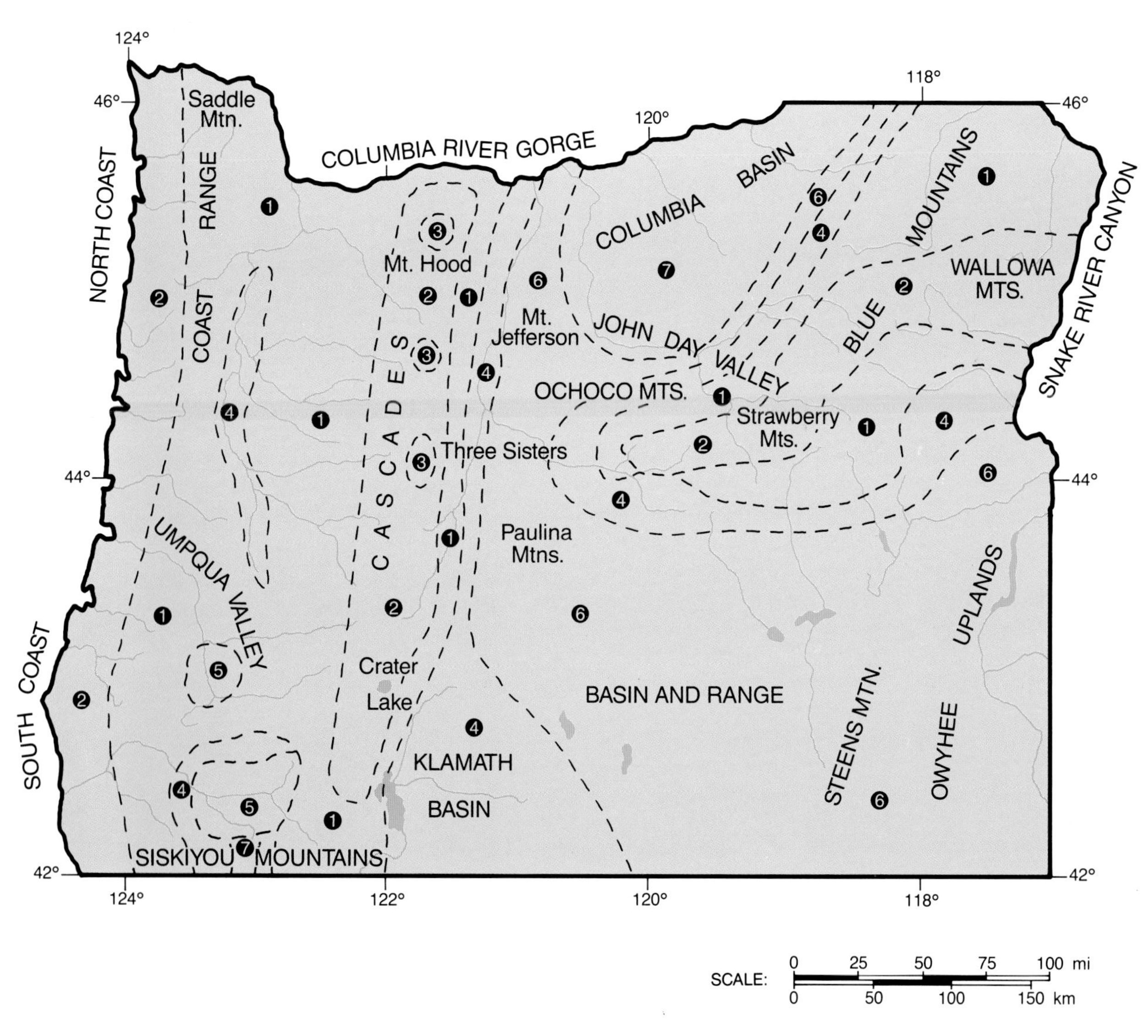

Major Vegetation Types After Detling, 1968

Boreal types:

1. Mesic Conifer Forest
2. Boreal Forest
3. Alpine Fell-fields

Austral types:

4. Pine-oak Forest
5. Chaparral
6. Juniper-sagebrush Woodland
7. Grasslands

Plant Names:

As this book is written for the amateur botanist as well as the professional, common names are given first in the text. Some plants have several common names. Generally the most frequently used name is given, sometimes more than one if they are particularly interesting or meaningful. Following the common name is the botanical name in parenthesis and italics, along with the abbreviations of the authors of those names. For each plant the genus, species, and, if applicable, sub-species and/or varietal names are given. The book is arranged in alphabetical order by genus, then species, sub- species, and varieties. For those who are interested, there is a list in Appendix II indicating the more scientific Englerian sequence of the plants included. This order shows better the relationship of plants with each other as to structure, genetics, and complexity of development. The plant family, with common name first, is given for each entry.

Plant Descriptions:

Generally, in describing the plant, only the most prominent features are given. For complete detailed description of each plant, the texts listed in the Bibliography are recommended. Most often data is given regarding the size of the plant and its component parts, with brief descriptions of stems, leaves, calyces, and corollas, the parts most people see first as they examine a plant. Attention is given to color, shape, texture, margins, and degree of hairiness and stickiness. Stamens, pistils, ovaries, bracts are discussed when important. Also, seed pods or fruits may be described and shown in the pictures where possible. Very little is said about underground parts since viewing these would result in a disturbance to the plant, even its loss. An attempt has been made to use words of normal usage in the descriptions, although this was not always possible. A glossary of some of the more technical terms may be found in Appendix III.

Plant Photography:

Every photographer has his own ways of doing things. Many different types of photographic equipment and supplies are on the market today. The author does not choose to recommend brand names of equipment or film, but will indicate what he has found to be successful for him, and make recommendations to those who would enjoy the pleasure of capturing, on film, some of the most beautiful creations found on our planet. Most of the photographs in this book were done with an Olympus SLR (through the lens viewing) camera. A 50mm macro, closing down to f22, has been the most useful and satisfactory type of lens. Other macros, including zoom types, have been used, but with less pleasing results. A 2X Macro Focusing Teleconverter has been used successfully in combination with the 50mm macro in achieving the extremely close-up photos. A synchronized flash attachment, Olympus T-32, has enabled the author not only to provide adequate light in difficult lighting conditions, and to catch flowers in breezy situations, but also to provide the greatest depth of field possible by closing the lens down to its smallest aperture. Use of a flash will often highlight the plant and play down the background, sometimes to the point of total blackness, removing all distracting objects, and in some situations lending an artistic quality to the picture. Some do not like to exclude the environment of the plant from the photo. Whatever your tastes, shoot the picture with and without flash, then choose the one you like best. Another danger of using flash is washout by too much light. Even though the flash is synchronized with the camera for exposure, it does not always provide optimum results. Bracket your shots by closing down one and two stops. It is seldom necessary to bracket by opening up the lens when you use flash in close-up situations. Advisably, the flash should be set at an angle to the object being photographed, to give the effect of side or overhead lighting, and to prevent harsh reflections from shiny parts of the plant. Choose the best flower specimen you can find; even better, use several different plants, as some flaw in one may not be noticed until later when you see your picture. Photograph the plant from several angles. Attempt to show or include all the prominent parts of the plant—all the marks of identification that are possible. Screen out objects that are extremely distracting, even though naturally there. The author has chosen in most cases to do some grooming of the surroundings of the plant, especially the removal of dry grass and light colored twigs that produce bright lines through the background, often appearing to grow out of the plant itself. Sometimes adjacent plants, if weeds or grass, can be removed, but do so carefully, so as not to disturb the roots or structure of the rare plant you are photographing. Sometimes it is better to hold an adjacent plant out of the picture by temporarily placing something on it or against it, or by having another person hold it for you. One should not shoot just the flower, as beautiful as it may be; other parts may be much more important in determining a plants identification. Shoot some close-ups and some pictures of the plant structure in its entirety. Sometimes separate shots of the leaves, seed pods or fruits are necessary. It probably will be necessary to return at another time to shoot fruits or seed pods. That could lead to a lot of travel if it is across the state from you! Side views, top views, and views from underneath the plant are important, though sometimes difficult. Refrain from picking flowers, seed pods, fruits, or other parts of the plant for easier photography, though it may be a very tempting thing to do. Also, and most importantly, be very careful about disturbing other plants in the area of your photo activity. Trampling or breaking down rare plants can possibly lead to their destruction or even extinction. It is good, when possible, to include insects that may be present on the plant. You may be showing the plant's most important pollinizers. Film is another choice that varies with the taste of the person using it. All that can be said here is that most available films have been tried, but the author always returns to Kodachrome 64, slide film, for its color reproduction, its lasting durability, its dependability, its adequate speed with relatively small grain size, and consequently its ability to be beautifully enlarged.

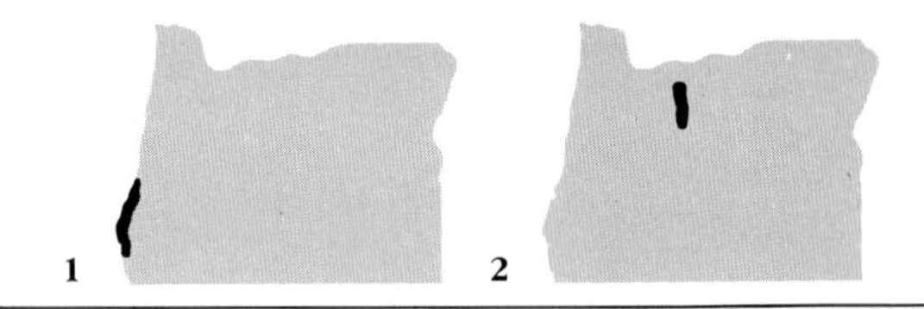

1
Pink sand verbena
(*Abronia umbellata Lam. ssp. breviflora [Standl.] Munz).*
Four-o'clock Family (Nyctaginaceae).
This extremely rare plant is known from only a few sites along the coast in southern Oregon and northern California. It grows in the soft sand above high tide line. This plant is very susceptible to disturbance and endangered throughout its range. The pictured flower was within inches of the track of a sand bike, narrowly escaping destruction.

Abronia umbellata ssp. breviflora is prostrate on the sand and is very glandular and puberulent, thus always heavily coated with particles of sand. The rosy-pink flowers form in round clusters at the top of a stem three to four inches tall. The flowers consist of sepals that appear to have wavy margins; there are no petals. The leaves are ovate to oblong, rounded at the tip and attached with stems longer than the leaf-blades. It blooms from May into autumn.

2
Tall agoseris
(*Agoseris elata* [Nutt.] Greene).
Composite Family (*Asteraceae*).
This rare species grows in meadows at mid-elevations in the Cascades of Oregon and Washington, and south into California. It is considered to be endangered in Oregon, and sensitive in Washington.

Also called a false dandelion, or mountain dandelion, because its flowering head, like true dandelions, consists of all ray flowers. *Agoseris elata* is a very large member of the genus *Agoseris*, with stems reaching twenty-six inches in height. Its bright yellow flower heads may be as large as one and one-half inches in diameter. It blooms from June to August.

3 4

3

Narrow-leaved water plantain

(*Alisma gramineum Gmel.* var. *angustissimum* [DC.] Hendricks).

Water Plantain Family (*Alismaceae*).

As the common name implies, this plant grows in water, sometimes in marshes, and on the edges of lakes and ponds. Though it grows natively most of the way across North America and Eurasia, in Oregon it is found only in a few scattered sites in the eastern part of the state from the Columbia River to Malheur County.

In this species and variety the flower scape is not longer than the leaves, and the leaves are narrower and stand stiffly erect, not usually submersed. The flower has three orbicular sepals and three white-pinkish petals, somewhat longer than the sepals, with smooth, unfringed margins. It may be found blooming throughout the summer.

4

Bolander's onion

(*Allium bolanderi* Wats.)

Lily Family (*Liliaceae*).

This wild onion of southwest Oregon is threatened in Oregon but more common in California. It grows in clayey soil, in gravelly areas, and sometimes on serpentine.

Allium bolanderi is about ten inches tall, with two leaves about the same height; its stem is topped by an umbel of six to twenty-five flowers. The tepals (petals and sepals) are similar in shape and color, and about equal in size. They are slightly less than one-half inch long, white to rosy pink with pink midribs, and sharply pointed. The inner petals may have finely toothed margins. The stamens are about one-half the length of the petals, and have pinkish anthers. It blooms from May to July.

5
Nevius' onion
(*Allium douglasii* Hook. *var. nevii* [Wats.] Ownbey & Mingrone).
Lily Family (*Liliaceae*).
This wild onion is endemic to a small area at the east end of the Columbia Gorge, in Hood River and Wasco counties in Oregon, and Klickitat County in Washington. It grows in moist grassy places which may be wet in the spring but later become dry. Its populations are considered stable in Oregon at present, but could become threatened in the foreseeable future.

Allium douglasii var. nevii is a very attractive onion with its tall stems (sixteen inches) topped with a ball-shaped umbel of deep pink to sometimes nearly white flowers. The stamens are equal in length to the tepals which are about one-third of an inch long. The stem, not thickened below the umbel, is longer than the leaves which are channeled; sometimes curved, sometimes straight. It blooms in May.

6
Rock onion
(*Allium macrum* Wats.).
Lily Family (*Liliaceae*).
This rare but currently stable onion may be found in dry gravelly soil, in the Blue Mountains of eastern Oregon, north into southeastern Washington and south to Klamath County.

Allium macrum is generally six to eight inches tall, exceeded in length by the narrow leaves. The sharply-tipped tepals are white to pale pink with distinct purplish midveins. The stamens, about equal to the tepals in length, have reddish-purple anthers. It blooms in April and May.

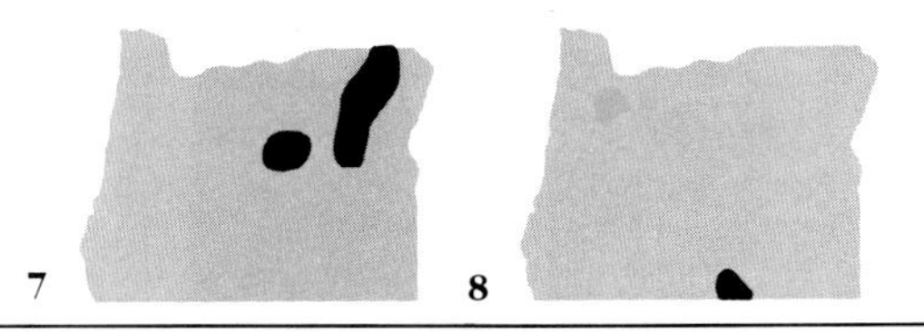

7
Swamp onion
(*Allium madidum* Wats.).
Lily Family (*Liliaceae*).
As its common name implies, swamp onion prefers wet areas in which to grow. It is found in the Blue and Wallowa mountains of eastern Oregon, and in Valley and Adams counties in Idaho. At present it is very limited in abundance and range, but currently stable.

The leaves of *Allium madidum* are generally shorter than the scape (flowering stem), about one-quarter of an inch thick, and channeled. Its scape may be ten inches tall. The flowers are white to pinkish and have tepals about one-third of an inch long with green midribs. The bulblets which form around the bulb distinguish it from similar species. It blooms in May and June.

8
Broad-stemmed onion
(*Allium platycaule* Wats.).
Lily Family (*Liliaceae*).
In Oregon, it is found only in southern Lake County. It is more common in the Sierra Nevada Mountains of California and Nevada. Though its range is small in Oregon it is considered to be currently stable.

Also called the Pink Star Onion, *Allium platycaule* has deep rose flowers shaped like stars. Each individual flower is constricted immediately above its ovary. The flower clusters are dense, numbering as many as thirty to ninety in one cluster. Both the leaves and the stem are broad and flattened. The leaves are falcate from their base, arching back toward the ground. It blooms in May.

9

Many-flowered onion

(*Allium pleianthum* Wats.).
Lily Family (*Liliaceae*).
Found only in the heavy, sticky clay of the John Day Valley in central Oregon, this rare endemic is limited in abundance throughout its narrow range but is currently stable.

The stem of *Allium pleianthum* is oblique to the bulb, flattened, and winged on both sides. It is short, usually less than four inches tall. The leaves, too, are flattened, generally falcate, and much longer than the scape. The tepals, one-half inch long, are pale pink to lavender, lance-shaped, with sharply pointed tips. It blooms in April.

10

One-leaved onion

(*Allium unifolium* Kell.).
Lily Family (*Liliaceae*).
This onion, common in California especially in the San Francisco Bay area, is indeed rare in Oregon where it is known from only one disjunct population along the Yamhill River southwest of Willamina near the Yamhill-Polk County line.

Allium unifolium is a misnamed onion as it normally has two, sometimes three or four leaves instead of one. Its flowering stem grows up to sixteen inches tall; its leaves are somewhat flattened and considerably shorter than the stem. The flowers are pink with tepals over one-half inch long, becoming papery when in fruit. It blooms in June.

11 12

11
Bog anemone
(*Anemone oregana* Gray *var. felix* [Peck] Hitchc.).
Buttercup Family (*Ranunculaceae*).
Known in Oregon from only four or five bogs in Lincoln and Polk counties, and from a few sites in the state of Washington, it is considered very rare in Oregon, but more plentiful in Washington.

The leaves of *Anemone oregana var. felix* are trifoliate. It has five to seven sepals that are white with purplish areas on the outside. It has no petals. There are more than sixty stamens which distinguish it from the species. It blooms early, starting in March, and sometimes lasting into June.

12
Waldo rock cress
(*Arabis aculeolata* Greene).
Mustard Family (*Brassicaceae*).
Found in serpentine soil in southwestern Oregon, chiefly in Josephine and Curry counties, extending into northern California, this plant is considered to be rare in both Oregon and California.

Arabis aculeolata grows to fifteen inches tall. It has a rosette of small (one-half to one inch long) basal leaves that are thickly covered by coarse simple and branched hairs. The stem leaves are also small, sessile, seeming to hug the stem, and less hairy than the basal leaves. The raceme is short, maybe six to eight lovely deep rose-purple, four-petaled flowers. Even the calyx is deep red. The seed pods (siliques) are straight and erect, up to two and one-half inches long. It blooms from April to June.

Basal leaves of Waldo rock cress

13 14

13
Brewer's rock cress

(*Arabis breweri* Wats.).
Mustard Family (*Brassicaceae*).
Found only in southern Jackson and Josephine counties, Oregon, and in northern California, it is presently considered to be secure within its narrow range.

Arabis breweri grows up to twelve inches tall, has basal leaves up to an inch in length, which are entire or few-toothed, and heavily ciliated with three-forked hairs. The petals are purplish-red, three-eighths of an inch long; the sepals, also purple, are about one-eighth of an inch long. It blooms from April to June.

14
Koehlers's rock cress

(*Arabis koehleri* Howell *var. koehleri*).
Mustard Family (*Brassicaceae*).
This very rare rock cress is known only from a few rocky cliff sites in the vicinity of Roseburg, where it is considered to be endangered throughout its range.

Arabis koehleri var. koehleri is an unusual rock cress in that it is a small shrub rather than an herb. It grows to about eight inches tall with the flower stems extending another six inches. The flowers have bright, purple-red flower petals, each about one-half inch long. It blooms early in April.

15 16

15
Koehler's stipitate rock cress
(*Arabis koehleri* Howell *var. stipitata* Rollins).
Mustard Family (*Brassicaceae*).
This variety is found in serpentine soil in Josephine and Curry counties in Oregon, and south into northern California. It is currently stable but could become threatened in Oregon in the foreseeable future. Though it is considered to be a variety of *Arabis koehleri*, it is morphologically quite different. It is not a shrub as is *Arabis koehleri var. koehleri*, although its base is somewhat shrubby.

Arabis koehleri var. stipitata, as the name implies, is "stipitate", meaning the pistil (and subsequently the fruit) is supported by a short stalk (stipe). The flowers are purplish-red, with petals that are one-half inch long, and siliques (fruits) that are three inches long, erect, or spreading. The basal leaves are numerous, and pubescent with star-like hairs. The upper portion of the stem is usually hairless. It blooms March and April.

16
McDonald's rock cress, also known as Red Mountain rock cress
(*Arabis macdonaldiana* Eastw.)
Mustard Family (*Brassicaceae*).
There is much confusion at this time as to the validity of this name for the species found in Curry County, Oregon. Some argue that the only true *Arabis macdonaldiana* is the one found on Red Mountain in California, and that the one in Oregon is a new species, simply being referred to as *Species novum*. *Arabis macdonaldiana* is listed Federally as endangered. *Species novum* in Oregon, whatever its name, is considered to be threatened throughout its range.

Characteristic of this plant are its rose-purple petals, each about three-eighths of an inch long, its spatulate basal leaves which are three-quarters of an inch in length, shiny, and hairless with one or two teeth on each margin, and its siliques that are about one and one-half inches long. It blooms in April.

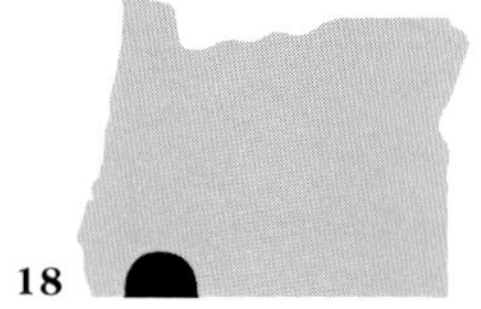

17
Rogue Canyon rock cress
(*Arabis modesta* Rollins).
Mustard Family (*Brassicaceae*).
This rare, possibly threatened rock cress is known only from the Rogue River canyon near Galice, Josephine County, Oregon, and perhaps a few isolated locations in California.

Arabis modesta is an herb growing to twenty inches tall, with numerous basal leaves that are stemmed and are up to two and one-half inches long. They are sometimes purplish beneath, covered with very small, appressed, four-parted hairs. The stem is hairless; the stem leaves (one inch long) are sessile but not auriculate. The flowers are pinkish-purple and have petals about one-half inch long. The fruits (siliques) are two inches long, ascending, and hairless. It may be found blooming in April and May.

18
California sandwort
(*Arenaria californica* [Gray] Brew.).
Pink Family (*Caryophyllaceae*).
Known from only a few sites in Jackson and Josephine counties, this sandwort is currently rare but stable in Oregon. It should be monitored occasionally. It is, however, far more common in California.

Arenaria californica is a very small, delicate, glabrous plant, one to four inches tall, with a thread-like stem, often branched near the base, and with minute paired leaves. The white flowers are usually single on a stem. The petals are longer than the sepals. It blooms from March to May.

19
Howell's sandwort
(*Arenaria howellii* Wats.; also *Minuartia howellii* Mattf.).
Pink Family (*Caryophyllaceae*).
This rare sandwort has been found only in Josephine and Curry counties in Oregon and Del Norte County, California. It prefers arid sandy and rocky soils.

Arenaria howellii is a very slender, freely branched plant up to fourteen inches tall. The leaves are mostly basal, those on the upper stems becoming bracts. The flowers are white, the petals somewhat longer than the sepals. It blooms from April into June.

20
Crater Lake sandwort
(*Arenaria pumicola* Cov.).
Pink Family (*Caryophyllaceae*).
This plant, also known as Pumice Sandwort as its species name suggests, is found in the volcanic soils of the Cascade Range. It is best known in the Pumice Desert of Crater Lake National Park.

The white petals of *Arenaria pumicola* are one-quarter inch long; its sepals are somewhat shorter and usually glandular and pubescent on the mid-ribs. The plant itself grows to about eight inches tall. Its leaves are filiform, nearly two inches long, and often ciliate on the margins. It blooms in July and August.

21
Prickly poppy
(*Argemone munita* Dur. & Hig. *ssp. rotundata* [Rydb.] G. Ownbey).
Poppy Family (*Papaveraceae*).
This plant, considered to be rare and threatened in Oregon, is known from the Alvord Desert region east of Steen's Mountain. It is more commonly found in the Great Basin areas to the south. It prefers dry, disturbed areas of pebbly or sandy soils.

Argemone munita ssp. rotundata is a large plant, growing to forty inches tall. It has six large white papery-like petals, from one inch to one and one-half inches long, surrounding a large, dense center of yellow stamens. There are numerous sharp spines on the stems and leaves. It blooms from June to August.

22
Clasping arnica or Streambank arnica
(*Arnica amplexicaulis* Nutt. *var. piperi* St. John & Warren).
Composite Family (*Asteraceae*).
This plant is found only along streams in the western part of the Columbia Gorge in Oregon and Washington. It has been reviewed for rarity in Oregon, but has not been listed because of taxonomic problems. It grades into the species outside the Columbia Gorge.

Arnica amplexicaulis var. piperi grows in large clusters, tall to twenty-four inches. It has four to six pairs of opposite, clasping stem leaves, which are glabrous, shiny, dark green, ovate, and often sharply and deeply toothed. The stem may be quite villous. Commonly it has three flower heads at the top of a stem, each head with eight to fourteen bright yellow ray flowers surrounding many yellow disk flowers. It blooms June to August.

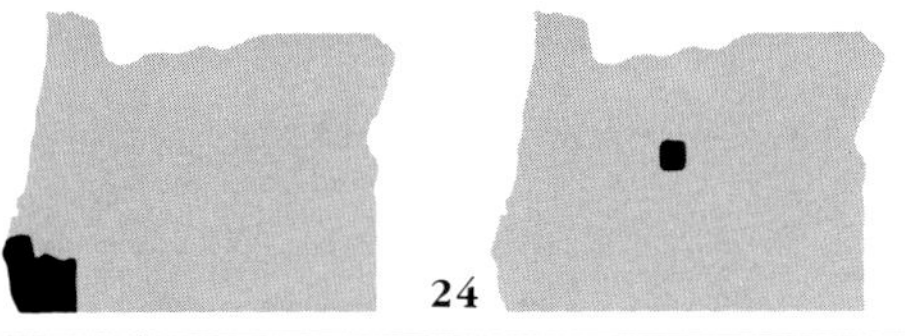

23
Nodding arnica or Serpentine arnica
(*Arnica cernua* Howell).
Composite Family (*Asteraceae*).
It is found on serpentine slopes in the mountains of Curry and Josephine counties and south into California where it is considered to be rare but stable.

The leaves of *Arnica cernua* are paired and opposite. The basal are on long petioles; the stem leaves are more narrow and not as deeply toothed as the basal. The plant grows to twelve inches tall, is sparsely ciliate, and usually has one flower head to a stem with about a dozen yellow ray flowers around a compact circle of yellow disk flowers. It blooms in April and May.

24
Estes' artemisia
(*Artemisia ludoviciana* Nutt. *ssp. estesii* Chamb.).
Composite Family (*Asteraceae*).
At one time known as *Artemisia douglasiana* Bess. *ssp. nomen ined.*, it has only recently been given its present name. This extremely rare, close relative to sagebrush (*Artemisia tridentata*), is known only from the Deschutes River near Cline Falls in Deschutes County in central Oregon and is endangered throughout its range.

Artemisia ludoviciana ssp. estesii grows four to five feet tall. Its leaves are entire, green, and somewhat glabrous on the upper surfaces, and woolly white underneath and on the stems. The flower racemes grow out of leaf axils, and the individual flower heads out of the axils of smaller leaves or bracts. The yellow, discoid flowers are sometimes obscurely radiate. It blooms from June to September.

Close-up of flower

25
Green-flowered wild-ginger
(*Asarum wagneri* Lu & Mesler).
Birthwort Family (*Aristolochiaceae*).
Only recently determined to be a separate and distinct species, it was formerly known as *Asarum caudatum* Lindl. *var. viridiflorum* Peck. Known originally only from the Lake-of-the-Woods and Mt. McLoughlin area in Klamath County, it has now been reported in Douglas, Josephine, and Curry counties in Oregon, where it is limited in abundance but currently stable throughout its range, which appears at present to be totally within the state. The plants grow in humus soils under a forest of red and/or white fir trees at moderately low elevations, up to open boulder fields at timberline.

Asarum wagneri is a perennial herb arising from elongated, horizontal rootstocks. Two deep green, kidney-shaped leaves that are somewhat wider than they are long, arise at each node. The leaves, two to three inches wide, have relatively long stems, that are covered with long, white woolly hairs. The leaves, too, are white woolly underneath, but shiny, glabrous on the upper surface except along some of the veins. The small flowers arise between the paired leaves on a stem that is much shorter than the leaf stems. Having no petals, the flower consists of three sepals, each about one inch long, that are bright green in the flared portion, turning a slight purplish color where the sepals bend into the calyx tube. The calyx lobes terminate in short, pointed tips that bend sharply upward. There are twelve stamens attached to the ovary, completely within the deep portion of the calyx tube. The fruit is a thick-walled capsule. It blooms from May into July.

26
Grass fern
(*Asplenium septentrionale* [L.] Hoffm.).
Fern Family (*Polypodiaceae*).
Although this is a wide-ranging species, it is rarely seen. Two small populations are known in the vicinity of the North Umpqua in Oregon. In California it is known to exist in only two highly restricted populations, three hundred miles apart, in Shasta and Tulare counties. It is considered endangered in both Oregon and California, but not mentioned in other adjacent states. It has also been reported from Utah, South Dakota, Oklahoma, and New Mexico, and is found in Europe and Asia. In Oregon it grows on the shady, moist, north faces of large rocks.

Asplenium septentrionale has green, grass-like fronds. The stipes (stems) of the frond are much longer than the blade (broader portion) which consists of two to three alternate, linear, lateral segments.

27
White-topped aster
(*Aster curtus* Cronq.).
Composite Family (*Asteraceae*).
This plant, which grew in the native grasslands once common from the Willamette Valley to Vancouver Island in British Columbia, has disappeared along with its habitat. Although somewhat more common in Washington, it is now known from only a few sites in Oregon, in Lane, Marion, and Multnomah counties, in small populations rediscovered since 1978. It is rare and threatened in Oregon.

Aster curtus is a small plant, four to twelve inches tall. It is often obscured by the herbaceous cover in which it grows, allowing it to be overlooked and possibly destroyed inadvertently. It is usually a single-stemmed plant, having numerous stem leaves with scabrous-ciliolate margins, alternately arranged up to the numerous flower heads which top the stem. The bracts surrounding each flower head are imbricate in three to four rows. Each head may have from one to three small, isolated, whitish ray flowers, and a few pale yellow disk flowers, some with purple anthers. The flowers bloom late into August and September.

28
Gorman's aster
(*Aster gormanii* [Piper] Blake).
Composite Family (*Asteraceae*).
This very rare, subalpine plant is found only on cliffs and in open rocky areas in the vicinity of Mt. Jefferson in the Oregon Cascade Mountains.

Aster gormanii is a small aster, growing four to twelve inches tall. It has glandular, sessile leaves, which are erect, crowded and overlapping up the stem. Its lower leaves are scale-like. The bracts of the flower head form an imbricate spiral. The flower head is usually single. There are ten to twelve white or pinkish ray flowers about one-half inch in length, surrounding the base of a column of yellow disk flowers. It blooms from July into September.

29
Wayside aster
(*Aster vialis* [Bradsh.] Blake).
Composite Family (*Asteraceae*).
This plant of open woodlands of the upper Willamette Valley was thought to be extinct until rediscovered in 1980 in the vicinity of Eugene. It is extremely rare throughout its limited range.

Aster vialis is a rayless aster, growing up to four feet tall. It has numerous long (two to three and one-half inches) lanceolate, sessile, somemtimes toothed leaves ranging the full length of the stem. The heads of yellowish disk flowers are rather few. There are no ray flowers. The accompanying photograph shows the flowers shortly after blooming (late August). The normal blooming period is from July to early August.

30
Applegate's milkvetch
(*Astragalus applegatei* Peck).
Legume Family (*Fabaceae*).
This plant, one of the rarest in Oregon, is truly endangered. It is known to exist only in one or two sites in Klamath County in southern Oregon. It is not known to exist in adjacent California. Its best population is found in an expanding industrial area of Klamath Falls. Its habitat is moist meadows.

Astragalus applegatei is a slender perennial, often decumbent, with stems to sixteen inches long, which have seven to eleven narrow, slightly strigose leaflets. The flowers, whitish to lilac in color, are small with petals only one-fourth of an inch long; the seed pods, up to one-half of an inch long, are spreading or declining, straight, and faintly mottled. It blooms from June to early August.

31
Hood River milkvetch
(*Astragalus hoodianus* How., also known as *Astragalus reventus* Gray *var. oxytropidoides* [M. E. Jones] C. L. Hitchc.).
Legume Family (*Fabaceae*).
This plant which grows only in grassy meadows above the Columbia River near Hood River and The Dalles, and into Washington, is limited in abundance throughout its range but is currently stable.

Astragalus hoodianus is readily recognized by its unusual manner of growth. It consists of erect, closely clustered stems, up to a foot tall, topped with a head of as many as fifteen flowers. The flowers are white, slightly greenish, about three-quarters of an inch in length. Its pods are nearly erect, pubescent, and three-quarters of an inch long. The leaves, too, grow in in an erect manner, but are shorter than the flowering stems, and have as many as thirty-five canescently strigose, linear leaflets, each about one-half inch long. It may be found in bloom as early as March and as late as June.

32 33

32
Pauper milkvetch
(*Astragalus misellus* Wats.).
Legume Family (*Fabaceae*).
This rare plant, also known as Scabland Milkvetch, and Watson's Dwarf Locoweed, grows in sagebrush and grasslands in the dry areas of the John Day country of central Oregon, and east and north into the Columbia River basin of southeast Washington.

Astragalus misellus is a small, spreading, sometimes decumbent plant, ten inches tall, with leaves half the length of the flowering stems. The seventeen to twenty-one leaflets are narrow and less than one-half inch long, pubescent on the underside, and smooth and glabrous above. The flowers, one-quarter inch long, are white to ochroleucous, sometimes tinged with purple. The pods are slightly under an inch in length, pendulous, somewhat curved upward, and with a stipe about equal to the calyx. It blooms May to July.

33
Peck's milkvetch
(*Astragalus peckii* Piper).
Legume Family (*Fabaceae*).
This milkvetch, thought to be extinct since 1950, was rediscovered recently in Deschutes County, and is now considered to be threatened throughout its range. It grows on dry sandy ground and in pumice.

Astragalus peckii is very small, with stems four to twelve inches long, lying close to the ground. The leaves, two to three inches long, are evenly pinnate with eight to twelve leaflets, each one-quarter of an inch long. The terminal leaflet is confluent with the leaf stem. The plant is covered with appressed white hairs. The few flowers are white to pale yellow. The banner is delicately lined with purple. The pods are one-quarter of an inch long, and deeply sulcate (grooved on the back). It blooms in June.

34 35

34
Long-leaf locoweed

(*Astragalus reventus* Gray *var. reventus*).
Legume Family (*Fabaceae*).
It is quite similar to Astragalus hoodianus, which at one time was called *Astragalus reventus var. oxytropidoides*. Another common name is "Revenant locoweed", meaning "to return ghost-like." This rare plant is found in dry areas of eastern Oregon and adjacent Washington and Idaho.

The numerous stems of *Astragalus reventus var. reventus* are stout and rigid, up to sixteen inches tall, and support six to fifteen yellowish-white flowers. The calyces have minute appressed black hairs. The leaves are nearly erect, about eight inches long with twenty-one to thirty-seven leaflets. Each of the leaflets is about one-half inch long, hairy beneath, glabrous above, falcate and folded. It blooms May to July.

35
Sterile milkvetch

(*Astragalus sterilis Barneby*).
Legume Family (*Fabaceae*).
This very rare milkvetch is found in the Owyhee River basin of extreme eastern Oregon, and southwestern Idaho. Because it seems to prefer growing only in a specific layer of volcanic ash found in those areas, is unable to tolerate disturbance, and as its name implies, has difficulty in setting fertile seed, this species is considered to be endangered throughout its range.

Astragalus sterilis is small, only six to eight inches tall, with stems rising singly, or possibly two or three together. The leaves are three to four inches long with six to eight small leaflets, each one-eighth to one-quarter inches long. There are only two to five white, fading to yellow,flowers on a stem. The pods are pendulous, inflated, and purple mottled. It blooms from May to July.

Close-up of flowers

36
Tygh Valley milkvetch
(*Astragalus tyghensis* Peck).
Legume Family (*Fabaceae*).
This species is endemic to Tygh Valley east of Mt. Hood, in Wasco County. It once was found in sagebrush and bunchgrass communities in dry rocky, sandy areas, but is now limited mostly to roadsides. It is threatened, if not endangered, throughout its range.

Astragalus tyghensis is a perennial, growing from six to twenty inches tall. It is densely villous hairy throughout. The pinnate leaves are up to six inches long with fifteen to twenty-one oval leaflets, each about one-half inch long. The stems terminate in flowering racemes of from twenty to forty flowers in a very tight arrangement. The calyx, about one-third of an inch long, is yellow, and has long, pointed, darker colored teeth. The flowers are pale yellow, nearly one-half inch long. The pods, about one-quarter of an inch long, are white, villous, and horizontal to declining. The flowers are in bloom late May to early July.

37
Silky balsamroot
(*Balsamorhiza sericea* Weber).
Composite Family (*Asteraceae*).
This plant has only recently been described as a new, distinct species. Endemic to serpentine soils in the Illinois River area from Eight-dollar Mountain to Oregon Mountain, and south into Siskiyou and Trinity counties in California, it is currently stable in both states, but could become threatened throughout its range in the foreseeable future.

Balsamorhiza sericea is characterized by its large leaves which are divided pinnately and are covered by dense, minute, shiny hairs giving it a silvery, silky appearance. The flower scapes are longer than the leaves. The involucre, almost an inch high, supports nine to eighteen yellow ray flowers and numerous yellow disk flowers. It blooms April to June.

38
Bensonia

(*Bensoniella oregana* [Abr. and Bacig.] Morton). Saxifrage Family (*Saxifragaceae*).

This species, found in the Siskiyou Mountains of southwest Oregon and in Humboldt County in California, is considered to be limited in abundance throughout its range but currently stable. It prefers wet meadows and moist streamside sites in Pre-Cretaceous metasedimentary rock at elevations above 4000 feet.

Bensonia oregana can be identified by its leaf shape which is broad and ovate, five to nine lobed, two to three inches long, and generally smooth and hairless. It is attached by a light green petiole, which has numerous long, white, shaggy hairs. The stems, also very hairy, grow to fourteen inches tall. The small white sepals and petals are only about one-sixteenth to one-eighth of an inch long, and the light green seeds forming in the hypanthium turn black toward maturity. It blooms in June and July.

Close-up of flowers

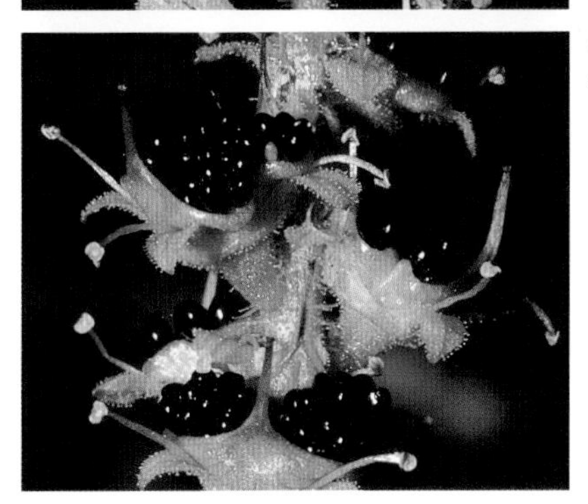

Close-up of mature seeds

39 40

39
Oregon bolandra
(*Bolandra oregana* Wats.).
Saxifrage Family (*Saxifragaceae*).
This species grows on wet cliffs in the Columbia Gorge, along the lower Willamette River, and in the Snake River Canyon. It is limited in abundance throughout its range, but currently stable.

The stems of *Bolandra oregana* come from bulblet-bearing rootstocks, and may grow to twenty-four inches in length. They are slender with numerous short, glandular hairs. The basal leaves are orbicular, deeply and sharply toothed. The three to four stem leaves are more like stipules. The flowers are few. The petals are purple, very narrow, sharply pointed, and up to one-half inch long. The calyx, yellowish-green and constricted at the neck like a vase, has long, sharp, spreading lobes. It blooms May to July.

40
Oregon grape-fern, or Pumice grape-fern
(*Botrychium pumicola* Cov. in Underw.)
Adder's-tongue Family (*Ophioglossaceae*).
One of the rarest of grape-ferns, it has been found only in pumice high on the volcanic peaks of Crater Lake, the Paulina Mountains, and Broken Top, in Oregon, and on Mt. Shasta in California. It is endangered throughout its range in both Oregon and California.

The stem of *Botrychium pumicola* is stout, grayish-green, about four to nine inches tall. It has both a sterile frond (leaf) and a fertile one. The sterile frond is erect, sessile, leathery, and about one to one and one-half inches long, having a glaucous or powdery-like surface appearance. It is usually ternately divided, each of the three sections pinnately sectioned into roundish segments. The fertile frond, taller than the sterile, is pinnately branched and carries the spore-bearing sporangia. They appear round and yellowish in color and contain a copious number of spores. The fertile frond is present from July to September.

41 42

41
Green's mariposa-lily
(*Calochortus greenei* Wats.).
Lily Family (*Liliaceae*).
This very rare flower grows on dry, brushy hillsides in southern Jackson County, Oregon, and in Siskiyou and Modoc counties in California. Two problems contribute to the rarity of this plant, 1) it is very palatable to grazing animals, and 2) it is attractive to collectors. These may lead to its extinction.

Calochortus greenei grows about twelve inches high, and has petals one and one-quarter inches long. Its petals are lilac on the outside, and banded at the base with yellow and deeper lilac. The inner surface of the petals are covered with dense white hairs, changing to yellow toward the base. The sepals, narrowly ovate, are slightly over an inch long, and are greenish with a tinge of purple. The anthers are large and obtuse. It has one basal leaf, about as tall as the flowering stem, and several linear, leaf-like bracts where the stem divides into two to five flower heads. It blooms in June and early July.

Inside view of flower

42
Howell's calochortus or Howell's mariposa
(*Calochortus howellii* Wats.).
Lily Family (*Liliaceae*).
A species threatened throughout its range, it is found in hot, dry wooded areas growing in serpentine rocky soil in Josephine County, Oregon, but apparently not ranging south into northern California.

Calochortus howellii grows up to eighteen inches tall. The flowers are a pure white shading to deep brown on the inside near the base. The inner side of the petals is densely covered with white, and near the base, dark brown hairs. The sepals, too, are white, much narrower than the petals and somewhat shorter. The leaves are solitary, slightly longer than the scape, and about one-quarter of an inch wide. Linear, leaf-like bracts subtend the branches to the flowering heads. It blooms in June and July.

43
Long-haired mariposa
(*Calochortus longebarbatus* Wats. *var. peckii Ownbey*).
Lily Family (*Liliaceae*).
This rare mariposa lily of the Ochoco Mountains in central Oregon is limited in abundance but currently stable. It prefers moist, grassy areas which become dry by mid summer.

Calochortus longebarbatus var. peckii is about ten inches tall. The corolla is campanulate, broad at the base. Its petals are from one to one and one-quarter inches long, and are lilac, with a deep purplish-red spot near the gland on the inner side. The gland is covered with woolly-yellow hairs, and there are long, white, curly hairs above the gland. The sepals are narrow, pointed, and greenish, dotted with red. It has a basal leaf nearly as tall as the flower stem, a cauline leaf arising from a bulblet near the base of the stem, and two narrow, opposite, lance-shaped bracts about two inches below the flowering head. It blooms in June and July.

44
Sego lily or Bruneau mariposa
(*Calochortus bruneaunis* Nels. & Macbr. This plant has also been known botanically as *Calochortus nuttallii* Torr. *var. bruneaunis* [Nels. & Macbr.] Ownb.)
Lily Family (*Liliaceae*).
It was first found in Bruneau Canyon in southern Idaho. Although its range extends from southern Harney and Malheur counties in Oregon to South Dakota, New Mexico, California, and Utah, where it is the state flower, it is rare in Oregon. It grows in dry sagebrush desert.

Calochortus nuttallii var. bruneaunis reaches twelve inches in height, and has petals one and one-quarter inches long. The petals are pure white with purple and yellow "eyes" at the base on the inner surface. They are abruptly pointed, and slightly recurved. The sepals are narrow, white, and shorter than the petals. It blooms from May to August.

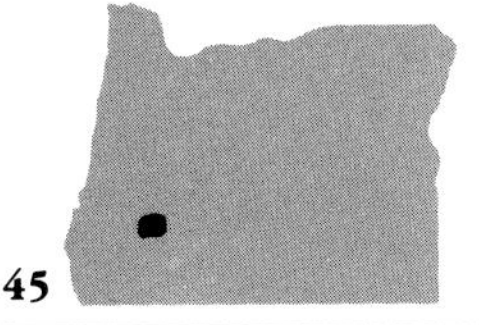

45
Umpqua mariposa-lily
(*Calochortus umpquaensis* Fredricks).
Lily Family (*Liliaceae*).
Originally known as the *Calochortus howellii* of Douglas County, Oregon, this newly-recognized species (description just recently published) is found only in serpentine rock and soil in a few small areas of the North Umpqua drainage. It is considered to be endangered throughout its range and is a candidate for federal listing.

The Umpqua mariposa grows about twelve to fifteen inches tall. The foliage is generally glabrous and glaucous. There is one basal leaf, as tall or slightly taller than the main stem, usually a smaller clasping leaf about one-third of the way up the stem, and pairs of lanceolate, leaf-like bracts about one inch long located at the point of division for each flower head. There are generally two to five large, showy flower heads on each plant. The petals are pure white, one and one-half inches in length, broad near the apex, and hairless on the outer surface which displays an arcuate green area near the base. The inner surface of the petals are extremely white-hairy from very near the shallowly toothed distal margins to the vicinity of the gland at the base. The gland is dark brown, and thickly covered with dark brown hairs. The sepals are narrow, pointed, and about three-quarters the length of the petals. The stigma is deeply triple-divided; the strongly three-ribbed, superior ovary becomes enlarged and pendulous in fruit. The six stamens have yellowish-colored, sharp-pointed anthers. It blooms from the latter part of May into June.

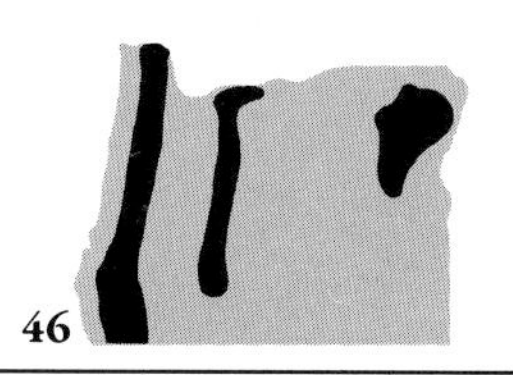

46
Fairy-slipper

(*Calypso bulbosa* [L.] Oakes.).
Orchid Family (*Orchidaceae*).

Reviewed for rarity and placed on a list of plants to be monitored because of its attractiveness to pickers and collectors, this is indeed a special plant, a treat to run across in the woods. It is an even greater treat to find the very rare *forma alba* of the species as is shown in the photograph. This orchid is found on precipitous slopes in the Columbia Gorge, on the west side of the Cascades, in the Coast Range, in the mountains of eastern Oregon, and as far south as San Francisco, California. It is said that, if picked, the bulb will die, as it tends to "bleed" to death. Neither will it take to transplantation.

Calypso bulbosa grows from a solid bulb up to ten inches in height, has a single, broadly oval, parallel-veined leaf near the base of the stem, and a vertical floral bract at the top of the stem. The stem is covered with two or three membranous sheathing bracts. Each plant has only one flower which is pinkish-purple in color. The lower lip of the flower, a slipper-like petal about three-quarters of an inch in length, is mottled with maroon and sometimes yellow or white. It usually has a patch of woolly, white hairs in the center and has two horn-like projections at the tip. The other petals and sepals are purple and are linear to lanceolate in shape. The column of stamens and style is petal-like, broad, oval, concave, with an anther near the tip, opening by a lid. It blooms in March in the low-lands, and later at higher elevations.

Albino form

47 48

47
Cusick's camas
(*Camassia cusickii* Wats.).
Lily Family (*Liliaceae*).
This tall camas is found only in the southeast portion of the Wallowa Mountains in Wallowa and Baker counties, Oregon, and in adjacent Idaho across the Snake River Canyon. It grows at elevations of 4500-5000 feet, on steep, rocky hillsides and in damp meadows, often associated with ponderosa pine. It is a regional endemic which needs to be monitored in both Oregon and Idaho.

Camassia cusickii is a large glabrous perennial, up to thirty inches tall, growing from a large bulb. The plants occur in thick clusters. The leaves, up to a dozen in number, are basal, one-half to two inches broad, twelve to twenty inches long, and folded along the center vein. The flowers, in a dense raceme ten to sixteen inches long, are a light purplish-blue. The tepals, about three-quarters of an inch long, are linear and pointed; the lowest tepal is somewhat remote from the others. They wither separately after blooming. It blooms from late April to late June.

48
White-flowered camas lily
(*Camassia leichtlinii* [Bak. J. G.] Wats. *var. leichtlinii*).
Lily Family (*Liliaceae*).
Found only in the Umpqua River basin near Roseburg and Sutherlin, this regional endemic can be seen in moist fields on either side of highway I-5, as the highway was put through the best population.

Camassia leichtlinii var. leichtlinii is a tall camas, reaching two feet in height. Unlike most camas it is white, not blue. The flowers, three-quarters of an inch to one and one-half inches long, are grouped in a raceme at the top of the stem, only a few of which are in bloom at a time. After blooming the petals dry and twist together over the ovary. There are several linear leaves which are keeled and somewhat shorter than the scape. It blooms in April and May.

49
Saddle Mountain bittercress
(*Cardamine pattersonii* Hend.).
Mustard Family (*Brassicaceae*).
This plant is considered to be endemic to Saddle Mountain, Onion Peak, and possibly two or three other peaks in the Coast Range of Clatsop County. It has been found nowhere else, and is considered to be threatened throughout its range.

Cardamine pattersonii may be seen in several places along the summit trail on Saddle Mountain, where its bright pink blossoms are quite showy. Each blossom is subtended by a bract, unique in this genus. Its leaves are a shiny deep green; each basal leaf has three to five leaflets, the terminal one is usually three-lobed. The plant grows four to eight inches tall; each flower is about one-half inch in diameter. It blooms in May and June.

Close-up of basal leaves

50
Willamette Valley bittercress
(*Cardamine penduliflora* Schulz.).
Mustard Family (*Brassicaceae*).
This bittercress is found only in the wet meadows and swamps in the Willamette Valley and west into the Oregon Coast Range.

Cardamine penduliflora grows six to sixteen inches tall. The stem leaves have five to nine leaflets; the terminal one, with three lobes, is much the largest. The basal leaves are larger, pinnate, long-stemmed, and sometimes have bulblets in their axils. The pure white flowers have four petals, each slightly less than one-half inch in length. The seed pods are erect siliques, one to two inches long. It blooms from April to May.

51 52

51
Green-tinged paintbrush
(*Castilleja chlorotica* Piper).
Figwort Family (*Scrophulariaceae*).
This rare plant, found only on Pine Mountain in Deschutes County and further south in Lake and Klamath counties, Oregon, is considered to be threatened throughout its range. It grows at rather high altitudes (6000 to 8000 feet) on dry gravelly or sandy slopes.

Castilleja chlorotica grows from four to fourteen inches tall and has glandular, long, soft hairs. Its lower leaves are linear, and entire; the upper leaves often have a pair of lateral lobes. The greenish-white flowers are inconspicuous, protruding from the inflorescence of leaves and bracts at the top of the stem. The lateral margins of the upper lip of the flower in this species are reddish to orange in color. The bracts are tipped with yellowish-green with some purple tones. It blooms late June to mid-August.

52
Slender indian paintbrush
(*Castilleja elata* Piper). Also known as *Castilleja miniata* Dougl. ex Hook. *ssp. elata* [Piper] (Munz).
Figwort Family (*Scrophulariaceae*).
This rare species of high meadows and bogs (up to 5000 feet), often on serpentine, is found only in Josephine and Curry counties in southwest Oregon and in Siskiyou and Del Norte counties in northern California.

Castilleja elata reaches a height of up to twenty-four inches. The upper bracts and calyces usually are tipped with a deep red color, though they sometimes vary from pink to a deep purplish-red. The flowers are up to an inch in length. It is not glandular-pubescent below its rather short inflorescence. It blooms from May to August.

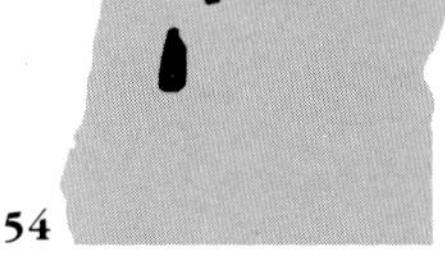

53
Glandular paintbrush, also called Sticky Indian paintbrush
(*Castilleja glandulifera* Penn.).
Figwort Family (*Scrophulariaceae*).
This regional endemic is found on gravelly soil high in the mountains of eastern Oregon (above 7500 feet in the Blue Mountains and on Strawberry Mountain).

As the name implies, glandular paintbrush is covered with sticky, glandular hairs. Its stem is stout, erect and angled, growing to sixteen inches high. Its leaves are usually entire, although the upper ones may have a pair of linear lateral lobes. The bracts and calyces are tipped with pale yellow, varying to a dull orange-red. The flowers, about an inch in length, are yellowish-green. It blooms July and August.

54
Golden paintbrush
(*Castilleja levisecta* Greenm.).
Figwort Family (*Scrophulariaceae*).
This yellow paintbrush of northwestern Oregon and western Washington was once quite common in the Willamette Valley in Linn, Marion, and Multnomah counties, but is now probably extinct in Oregon. A few sites still exist in Washington in Clark and Thurston counties where it is considered endangered.

Castilleja levisecta is the only yellow-bracted Indian paintbrush in its Willamette Valley-Puget Trough range. It grows from a perennial base to over twelve inches high, and is covered with a soft pubescence. Its leaves, closely ascending to the stalk, are narrowly oblong with one to four pair of short lobes near the tip. They are about one to one and one-half inches long, are closely ascending to the stalk, and turn reddish with age. The bright yellow bracts, also turning to reddish-orange with age, are oblong with one to two pairs of short lobes. The flowers barely extend beyond the bracts. It blooms from April through August.

55 56

55
Dixon's paintbrush
(*Castilleja miniata* Dougl. *var. dixonii* [Fern.] Nels. & Macbr., also known as *Castilleja dixonii* Fern.).
Figwort Family (*Scrophulariaceae*).
A species of the coast of Washington to southeast Alaska, it is known in Oregon only from the summit of Saddle Mountain in Clatsop County.

The stems of *Castilleja miniata var. dixonii* are solitary, curved to erect, and from ten to thirty inches tall. The lower parts are glabrous; the inflorescence is somewhat villous. The leaves are smooth-margined and lance-shaped. The bracts may have narrow lateral lobes. Both the bracts and the calyces are tipped with scarlet. The corolla is over an inch long; its upper lip or galea has narrow, pale or reddish margins. It blooms in late summer on Saddle Mountain.

56
Steens Mountain paintbrush
(*Castilleja pilosa* [Wats.] Rydg. *var. steenensis* [Penn.] N. Holmg., also known as *Castilleja steenensis* Penn.).
Figwort Family (*Scrophulariaceae*).
To date this variety of *Castilleja pilosa* has been found only at high elevations along the summit ridge on Steens Mountain, with one reported siting in the Hart Mountain National Wildlife Refuge in Lake County. Although locally abundant on Steens Mountain, its range is indeed small. It grows with grasses and other plants in rocky soil.

Castilleja pilosa var. steenensis is a small perennial reaching ten to twelve inches in height. Its stems and leaves are a reddish-green changing into a lighter, slightly yellowish-pink tone in the inflorescence. Its finely pubescent leaves are linear-lanceolate with one or two pairs of lobes. The bracts are broader with one or two pairs of lobes, and the calyces, which are nearly an inch long, are cleft, with each division being divided into two lobes. It blooms late June to early August.

57
Cliff paintbrush
(*Castilleja rupicola* Piper).
Figwort Family (*Scrophulariaceae*).
Generally a species of high elevations in the Cascade Mountains from central Oregon to British Columbia, it is found in the Columbia Gorge as low as four hundred feet. More information is needed to determine its degree of rarity in Oregon.

Castilleja rupicola grows in clumps. It is pubescent on the stems and leaves, becoming villous in the inflorescence. The leaves are oblong with two to three spreading linear lobes on each side. The many lobed bracts and cleft calyces are tinged a scarlet-red distally. The corolla is about an inch long. It blooms June to August in the mountains, but as early as April in the Columbia Gorge.

58
Yellow-haired paintbrush
(*Castilleja xanthrotricha* Penn.).
Figwort Family (*Scrophulariaceae*).
This plant is endemic to the John Day River area in central Oregon where it grows on dry rocky slopes in sagebrush country up to 2500 feet elevation. It is currently stable but could become threatened or endangered in the future, and therefore should be monitored.

The stems of *Castilleja xanthrotricha* are clustered, unbranched, ten inches tall, pubescent and somewhat glandular. The leaves are entire, one to two inches long. Its bracts are shorter and wider than the leaves and are tipped with a creamy-yellow to pinkish-brown color. The corolla may be over an inch in length. It blooms April to July.

59
Thick-stemmed wild cabbage
(*Caulanthus crassicaulis* [Torr.] Wats.).
Mustard Family (*Brassicaceae*).
This plant of the desert plains and lower mountains of the Great Basin of Nevada, central California, southern Idaho, and Utah is found just inside the Oregon border in southern Lake, Harney, and Malheur counties. It is rare in Oregon, but is much more common elsewhere.

Caulanthus crassicaulis is a strange appearing plant with a hollow, inflated, nearly leafless main stem, growing to a height of from twelve to forty inches. It is without hairs, but has a whitish powdery appearing surface. The basal leaves form a tight rosette around the stem. The lower ones are oblanceolate and almost entire; those above divided into deeply irregular segments. A loose raceme of flowers form along the upper portion of the broad stem. Each individual flower has a short, stout stem. The pink sepals, four in number, and over one-half inch long, are joined together to form an urn-shaped tube widest at the base. They are densely hirsute with white hairs. The petals, also four, are brownish-purple, longer than the sepals, and channeled. The two-lobed stigma and the stamens are not exserted beyond the inflorescence. The stout seed pods are erect, and about four to five inches long. It blooms from late May into July.

Close-up of flowers

60
Hoary chaenactis

(*Chaenactis douglasii* [Hook.] H. & A. *var. glandulosa* Cronq.).
Composite Family (*Asteraceae*).
This rarely occurring plant of northeast Oregon, southeast Washington and nearby parts of Idaho, is now considered stable throughout its range. It grows in pumice and in dry rocky or sandy places.

The stems of *Chaenactis douglasii var. glandulosa* are few or solitary, up to two feet tall, and have leaves up to four inches long which are pinnate or bipinnate and densely pubescent, though greener in this variety than in the species. In this variety the flowers are pink. Each flower head is about one-half inch tall with forty to forty-five disk flowers and no ray flowers. It blooms in June and July, and into September at higher elevations.

Basal leaves

61
John Day chaenactis

(*Chaenactis nevii* Gray).
Composite Family (*Asteraceae*).
Also known as Nevius' chaenactis, it is the only yellow-flowered chaenactis in our area. It is endemic to Wasco, Wheeler, and Grant counties in central Oregon. It is best known in the Painted Hills area, growing in the red and gray clay outcrops. It is locally abundant within a very narrow range.

Chaenactis nevii is an annual growing to twelve inches tall. Its leaves are pinnatifid, the lower ones are one and one-half inches long; the upper ones much smaller. The stems are usually branched, each branch terminating in a yellow, discoid (rayless) flower. The plant is puberulent and glandular. In dry years they are very scarce; in normal wet springs they can become so numerous as to color entire hillsides. It blooms May and June.

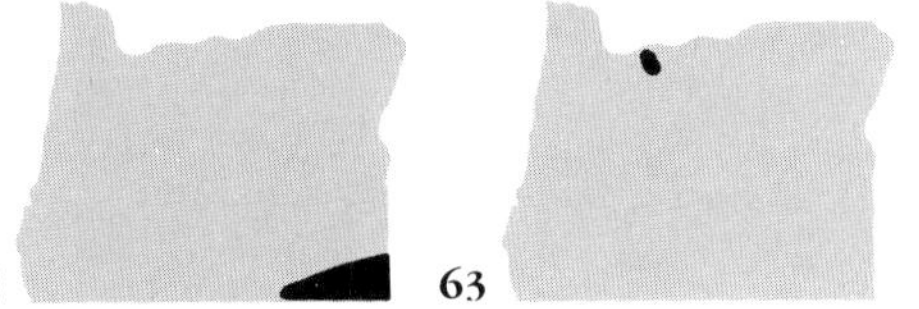

62
Broad-flowered chaenactis
(*Chaenactis stevioides* H. & A.).
Composite Family (*Asteraceae*).
This plant is known in Oregon only in Harney and Malheur counties. It is considered to be threatened in this state, but more common elsewhere. Its range extends east into Wyoming and south into New Mexico and California. It grows in clayey soil.

Chaenactis stevioides is a plant growing from four to twelve inches tall. Its leaves are simple-pinnate; the stems are slender and glandular. The disk flowers are white, sometimes tinged with rose. There are no ray flowers. It blooms in June.

63
Mt. Hood bugbane
(*Cimicifuga laciniata* Wats.).
Buttercup Family (*Ranunculaceae*).
This regional endemic, once known only from Lost Lake near Mt. Hood, has now been found from the Cascades near Mt. Hood, Oregon, to Silver Star Mountain in Washington. It grows in wet, rocky woodlands.

Cimicifuga laciniata reaches a height of five feet. Its leaves are laciniate, sharply and unevenly toothed, closely resembling baneberry (*Actea rubra*), but it blooms much later. The stem is glabrous up to the tomentose flower clusters. The flowers are white, and usually have petals (species *elata* does not). It blooms in August and September.

64
Nelson's thistle

(*Cirsium acanthodontum* Blake).
Composite Family (*Asteraceae*).
This thistle grows in the Rogue River area and on the coast south of the Rogue. It has also been reported on rare occasions further up the Oregon coast, and south into Humboldt County, California. The plant shown here was photographed on Cape Sebastian.

On the coastal headlands *Cirsium acanthodontum* is low and spreading, but inland it grows tall, up to forty inches. The basal leaves are narrowly ovate, pinnatifid, with prickly-ciliate margins, and a rather soft sharp spine at the tip of each lobe. They are woolly beneath and glabrous above. The upper leaves are similar, but clasping, without stems. The flower head is an inch or so tall; the bracts are tipped with spines and have prickly-ciliate margins. The flowers are pale pink to reddish-purple. It blooms from June to September.

65
Steens Mountain thistle

(*Cirsium peckii* Hend.).
Composite Family (*Asteraceae*).
This regional endemic, found only on Steens Mountain and in the Pueblo Mountains to the south, has a narrow range but is abundant where it does occur. It grows on dry slopes, along the canyon rims, and along roads and streams, from 5000 to 7500 feet.

Cirsium peckii is robust, up to five feet tall, with large flower heads clustered near the top, each two inches to two and one-half inches across. The heads are woolly with web-like hairs; the leaves and bracts have long, sharp, yellow spines. It blooms in July and August.

66
Copper bush
(*Cladothamnus pyrolaeflorus* Bong.).
Heath Family (*Ericaceae*).
This deciduous shrub of the north is found in a few scattered populations in damp forests at higher elevations in the Oregon Coast Range in Clatsop and Tillamook counties.

Cladothamnus pyrolaeflorus may grow up to six feet tall. It is an attractive bush with shiny dark green leaves and copper-colored flowers that are nearly an inch in diameter. The flowers have five sepals, five petals, ten stamens, and one long, recurving style, tipped with a disk-shaped stigma atop a bright green, superior ovary. You may expect to find it in bloom around the 4th of July.

67
Sierra spring beauty
(*Claytonia nevadensis* S. Wats.).
Purslane Family (*Portulacaceae*).
This species of the California Sierras is known in Oregon only from the Steens Mountain area where it grows in wet gravelly soils at high elevations below springs or melting snow.

Claytonia nevadensis is rhizomatous, growing in long continuous patches. It is perennial with reddish stems three to five inches long. The leaves are thick and fleshy, ovate, obtuse, green, and often with reddish margins. The flower petals are white to pink with deeper pink veins, green and clawed at the base. The filaments of the stamens are purple on the distal one-third; the anthers are a bright pinkish-purple. It blooms in August and September.

68 69

68
Umbellate spring beauty
(*Claytonia umbellata* Wats.).
Purslane Family (*Portulacaceae*).
Although widespread in California and Nevada, this plant is indeed uncommon in Oregon. It can be found in small separated sites in Deschutes, Wasco, Harney, and Wallowa counties. It is considered to be "limited in abundance, but currently stable in Oregon." It seems to prefer growing in a bed of small stones.

Claytonia umbellata is a small perennial growing from a corm on a long stem mostly underground. The leaves are thick and nearly oval. The flowers have two obtuse, pinkish sepals less than a quarter inch long, and five white to deep pink petals, which are ovate, entire, and half again as long as the sepals. It blooms in June and July.

69
Columbia virgin's-bower
(*Clematis columbiana* [Nutt.] Tor. & Gray *var. columbiana*).
Buttercup Family (*Ranunculaceae*).
This climbing plant, found growing over brush in open to deep woods, at mid-elevations, from the Blue and Wallowa mountains of northeast Oregon to Canada, is rare in Oregon.

The leaves of *Clematis columbiana var. columbiana* are divided into three ovate, entire or very slightly toothed leaflets, that are pointed at the tip, and somewhat heart-shaped at the base. The lateral leaflets are oblique or unsymmetrical. The leaves are hairy underneath but quite glabrous on the upper surface. The flowers, nodding and single on a stem, have purplish-blue sepals and no petals. The sepals are an inch to an inch and a half long, lance-shaped and pointed at the tip. This *Clematis* blooms from June to August.

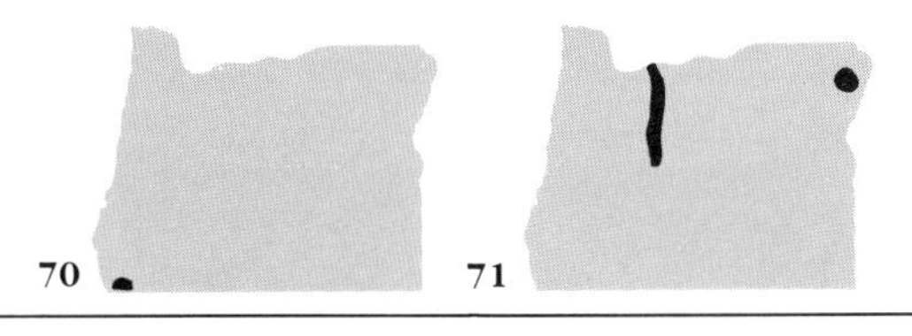

70
Red clintonia, also called Andrew's bead-lily
(*Clintonia andrewsiana* Torr.).
Lily Family (*Liliaceae*).
This California species follows the redwoods north to their northern limit just across the Oregon line near Brookings. It is very rare and endangered in Oregon but more common in California. There is some question as to whether it still exists in Oregon.

Clintonia andrewsiana grows twelve to twenty inches tall on a stout, erect stem. The showy flowers are pink to deep red, one-half to three-quarters inch long, with yellow anthers. There are four or five basal leaves that are elliptic in shape, six to twelve inches long, shiny green, and glabrous except for a few hairs along the margins and basal portion of the mid-rib. The berries are a metallic blue and less than a half inch long. It blooms May to July.

71
Larsen's collomia
(*Collomia debilis* [S. Wats.] Greene *var. larsenii* [Gray] Brand.).
Phlox Family (*Polemoniaceae*).
Also called Talus collomia because it grows on talus slopes on the high peaks of the Cascades. It is rare throughout its range from Washington to California.

Collomia debilis var. larsenii is a perennial growing from a very deep taproot. Its stems are only three to four inches tall and, along with its leaves, are quite puberulent and glandular. The leaves are less than an inch long, three-parted, and deeply lobed. The blossoms may be purple to pale pink with purple veins. The stamens are somewhat unequal in length. It blooms from July to September.

72
Bristle-flowered collomia
(*Collomia macrocalyx* Leib. ex Brand.).
Phlox Family (*Polemoniaceae*).
This is a plant known only from eastern Oregon where it grows in dry, rocky, undisturbed areas. With very few widely-separated populations still in existence, it is considered to be very rare but presently stable.

Collomia macrocalyx grows about four inches tall. Its foliage is covered with fine, white hairs, but it is not glandular. The principal leaves are clustered near the inflorescence. They are linear to lanceolate in shape, roughish, about an inch long, dilated and lighter colored near the base. The calyx is about four-tenths of an inch long with narrow and sharply-tipped sepals that are unequal in length. The funnel-shaped, five-lobed corolla is purple, turning to white in the throat. It blooms from late May to early June.

73
Mt. Mazama collomia
(*Collomia mazama* Coville. It is also known as *Gilia mazama* Nelson & Macbride).
Phlox Family (*Polemoniaceae*).
This rare species, limited in abundance and distribution, is endemic to a small area of the southern Oregon Cascades in Douglas, Jackson, and Klamath counties. It prefers dry woods at rather high elevations.

Collomia mazama, about ten inches tall, is glabrous below and has long, glandular, soft, white hairs in the inflorescence. The leaves are nearly two inches long and glabrous; the upper ones are coursely toothed. The flowers vary from pink to purple and blue; they are funnel-shaped and over one-half of an inch long. The stamens are exserted and have white anthers. It blooms in July and August.

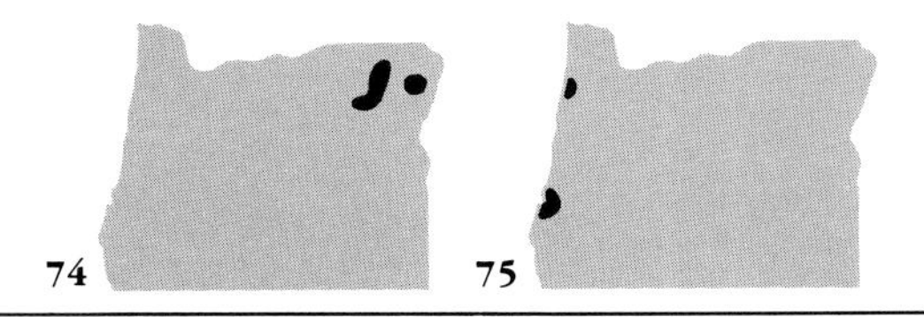

74
Yellow coral-root, or Early coral-root
(*Corallorhiza trifida* Chat.).
Orchid Family (*Orchidaceae*).
This plant was also known for a short time, 1880 to 1883, as *Corallorhiza corallorhiza*. In Oregon, Yellow Coral-root is found chiefly in the Blue and Wallowa mountains of the northeast corner. However there are a few reports of it in western Oregon where it is very rare. It is also native in Europe. It belongs to a group of mycotrophic plants that do not produce chlorophyl, but depend on the fungi in the soil for food. It prefers moist shaded areas on forest floors.

The yellow stem of *Corallorhiza trifida* reaches twelve inches in height; it has no leaves, merely scale-like bracts. The inflorescence on the upper part of the scape is loosely ten to twelve flowered. The petals and sepals are yellowish except for the lip which is an unspotted white. It blooms early in May and June.

75
Saltmarsh birdsbeak
(*Cordylanthus maritimus* Nutt. ex Benth. *ssp. palustris* [Behr] Chuang & Heckard).
Figwort Family (*Scrophulariaceae*).
This inconspicuous plant was difficult to see in the grasses along the shore of Coos Bay. It has been found in only one other site in Oregon, in Tillamook County, and in California where it is also disappearing and endangered. It grows in salt marshes, just above high tide line where it is greatly endangered by the draining, filling, and polluting of its habitat. Like Indian paintbrush, it is partially parasitic on other plants for its nutrients.

The corolla of *Cordylanthus maritimus ssp. palustris* is less than an inch long, and is usually pinkish to purple, though some of a yellowish-white color may be seen. The foliage is grayish-green, often villous. The floral bracts are oblong with a pair of short teeth at the tip. It is an annual that grows from four to twelve inches tall. It blooms from June to October.

76

Cold water corydalis

(*Corydalis aquae-gelidae* Peck & Wilson).
Bleeding Heart Family (*Fumariaceae*).
This rare plant is known from only a few populations in the drainage of the Clackamas River, in the Columbia River Gorge at fairly high elevations (up to 4000 feet), and in Skamania County Washington where it is also considered to be threatened. As its name implies, it prefers to be near cold water springs or streams.

Corydalis aquae-gelidae grows in erect clusters, to four feet tall, rooting from basal branches. Its leaves are four-parted pinnate. The stems end in a raceme of up to forty flowers. The flowers are white to cream colored, tipped with pink or a rosy lavender. The corolla is about three-quarters inch long; the upper petal is a strongly keeled hood which extends back to a spur about equal in length to the blade. It blooms in June and July.

77

Cushion cactus or Ball cactus

(*Coryphantha vivipara* [Nutt.] Britt. & Brown).
Cactus Family (*Cactaceae*).
One of only four cactus species known to occur in Oregon, it grows chiefly in the dry areas of the central part of the state. Although the range of this cactus extends east to Alberta, Colorado, western Kansas and Oklahoma, it is currently being reviewed for rarity in Oregon.

The plants of *Coryphantha vivipara* are subglobose, from one to five inches tall, covered with small nipplelike projections. It is sometimes referred to as the Beehive Nipple Cactus. From the top of these bumps protrude a swirl of three to five long central spines, and around them another ten to fifteen smaller ones. The pictured specimen is barely coming into bloom so doesn't show the rich pink to reddish-purple flowers. It blooms in June.

78
Baker's hawksbeard
(*Crepis bakeri* Greene *ssp. cusickii* [Eastw.] Babc. & Stebb.).
Composite Family (*Asteraceae*).
This is a somewhat rare species growing on arid rocky slopes and flats in Jackson, Klamath, and Lake counties of southern Oregon.

Crepis bakeri ssp. cusickii is a perennial plant, ten to twenty inches tall, with reddish, pubescent, and sometimes glandular stems. The leaves are broadly oblong-shaped, and are deeply parted into oblong or linear segments, that are generally entire along their margins. The basal leaves are about eight inches long, the upper ones much smaller. All the leaves are grayish-green with a pubescence. The flowers are few at the top of stems which widen near the inflorescence. There are ten to twelve pointed bracts surrounding the flower head. They are in two ranks, the outer ones fewer and much shorter than the inner ones. The corollas, about three-quarters of an inch long, are yellow and consist only of ligules (strap-like ray flowers). It blooms in May and June.

79
Baker's cypress, also known as Matthew's or Siskiyou cypress
(*Cupressus bakeri* Jeps. *ssp. matthewsii* Wolf).
Cypress Family (*Cupressaceae*).
This tree is found at scattered sites on dry wooded slopes, usually in serpentine soil, in Curry, Josephine, and Jackson counties in southern Oregon, and in Siskiyou County, California. It is rare in both states and endangered in Oregon.

This rather small cypress grows only to about sixty-five feet high with a trunk never exceeding twenty-four inches in diameter. The bark is thin, fibrous, rather grayish brown. Its branches are spreading and slender. The leaves are flat and strongly appressed, constricted between the pairs, with resin glands on the outer surface that are quite conspicuous and very aromatic. The male cones are short and round; the female cones are globe-shaped, shiny brown, and about one-half to three-quarters of an inch in diameter when mature.

Close-up of leaves and cone

80 81

80

Hayden's cymopterus

(*Cymopterus bipinnatus* Wats.).
Parsley Family (*Apiaceae*).
This species is found on high, dry, rocky ridges in central Oregon, and on the summit ridges of Steens Mountain. It is rare and threatened in Oregon but more common in the eastern part of its range in Idaho, Montana, Nevada, and Utah.

Cymopterus bipinnatus is a plant of numerous clustered stems that are up to ten inches in height. The leaves grow separately from the base, and are narrowly oblong, pinnately lobed, and roughly puberulent with a grayish-green color. The flowering stems are much longer than the leaf petioles. They bear a single umble of white flowers which is subtended by linear, acute bractlets that are irregular and scarious. The fruit is elliptic and strongly winged. This species blooms in May to June; later at higher elevations.

81

Yellow lady's slipper

(*Cypripedium calceolus* L. *var. parviflorum* [Salisb.] Fern.).
Orchid Family (*Orchidaceae*).
Once found in Oregon in the Warner Mountains of Lake County, and on Pea-vine Mountain in Josephine County, this species is now apparently extinct in this state. It is listed as endangered in Washington where only a few plants have been reported from two eastern counties. Its range extends into British Columbia, Idaho, Colorado, to eastern United States and Canada, and Europe and Asia.

The sepals and petals of *Cypripedium calceolus var. parviflorum* are brownish, except for the bright yellow lip (slipper) which has reddish-colored lines interiorly, and is up to one and one-half inches in length. The sterile stamen is yellow, dotted with reddish-purple. The plant grows twelve to twenty four inches tall, and has several alternate, oval leaves (five inches long) along the stem. It blooms from May to July.

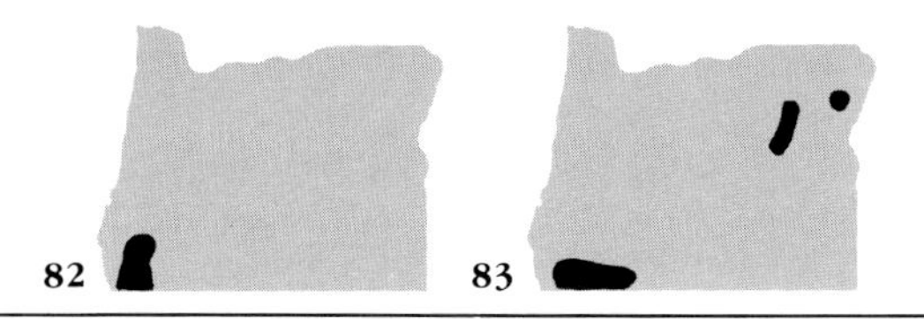

82
California lady's slipper
(*Cypripedium californicum* Gray).
Orchid Family (*Orchidaceae*).
Endemic to the Siskiyou Mountains of southwestern Oregon and northwestern California, this rare orchid has the most limited range of the lady's slippers known in Oregon. It grows in serpentine bogs, usually associated with *Darlingtonia californica*.

Cypripedium californicum grows twelve to twenty-four inches tall, is puberulent, has several clasping, ovate to lanceolate or oblong leaves (three to six inches long) extending the full length of the stem, and three to eight flowers arising singly from the axils of the leaf-like bracts on the upper part of the stem. The rounded, yellow sepals are over one-half inch in length; the white lower lip or "slipper" is nearly an inch long. It blooms in May.

83
Clustered lady's slipper
(*Cypripedium fasciculatum* Kell.).
Orchid Family (*Orchidaceae*).
This small rare orchid is known only from scattered populations in Oregon, Washington, California, Idaho, and Utah. Its habitat is in open woods of coniferous forests.

Cypripedium fasciculatum grows four to ten inches tall. The stems are woolly, and glandular; the two ovate leaves are opposite, sessile, and hairy underneath. Its sepals and lateral petals are about three-quarters of an inch long, greenish-brown to purple, and broadly lanceolate, often closely covering the lip. The slipper-like lip, one-half inch in length, is greenish-yellow with brownish-purple veins. The white sterile stamen is oblong, obtuse, and about the same length as the stigma. It blooms from April to July.

Close-up of flowers

84
Mountain lady's slipper
(*Cypripedium montanum* Dougl. ex Lindl.).
Orchid Family (*Orchidaceae*).
Once listed as very endangered in Oregon, this orchid has been found to still be relatively stable in the Blue Mountains although it has virtually disappeared from most of its former habitats in Oregon. It could become threatened throughout its range in the foreseeable future, however, as it tolerates no disturbance of its habitat and is attractive to pickers and diggers. In Oregon it can still be found at scattered sites on both sides of the Cascades, in the Siskiyous, the Ochocos, and the Wallowas. It now grows mainly in the moist woods of the mountains, occasionally side by side with clustered lady's slipper.

Cypripedium montanum is one of the larger lady's slippers. Reaching two feet in height, its stems are stout and lightly puberulent, with several dark green, broadly lance-shaped leaves, which are alternate and clasping. The sepals, nearly three inches long, are dark brown, narrow and twisted. The slipper petal or lip, over one inch long, is pure white with purple veins. The showy sterile stamen is yellow with purple spots. It blooms from May to August.

Leaf form

85
California pitcher-plant
(*Darlingtonia californica* Torr.).
Pitcher-plant Family (*Sarraceniaceae*).
It is also known as Cobra-lily due to the shape of its leaves. Though numerous at sites where it grows in coastal and inland bogs, from Tillamook County south into California, and from sea-level to 4000 feet, this plant has been considered threatened by collection for sale as a house plant.

As *Darlingtonia californica* depends in part for its sustenance on the ingestion of insects, it has highly modified leaves up to twenty inches tall that are tubular and hooded at the top. Below the hood is an opening which is bordered in front by a pair of prominent appendages. Small insects are attracted through this opening into the hollow leaf, from which few are able to escape because of the down-angled, glandular cilia lining the inside. The insects eventually fall into the water at the base of the leaf and are digested. The plant also produces chlorophyll. The stems, up to forty inches tall, are topped with a nodding flower having five yellow sepals that are one and one-half to two and one-half inches long, and five purple petals that are about two-thirds as long as the sepals. These surround a large superior ovary. As the plants go to seed, the developing capsule returns to an erect position. This plant blooms in May and June.

86 87

86
White rock larkspur
(*Delphinium leucophaeum* Greene).
Buttercup Family (*Ranunculaceae*).
This very rare larkspur is found in Oregon only in Clackamas, Marion, Multnomah, and Yamhill counties. It grows on cliffs and ledges in a few sites along the lower Willamette and Columbia Rivers and on the palisades above Lake Oswego. It has also been reported in Lewis and Klickitat counties in Washington. A narrow regional endemic, it is threatened throughout its range.

Growing from twelve to thirty inches tall, *Delphinium leucophaeum* has angled and flexuose stems. It is mostly glabrous, and topped with a many-flowered raceme. The sepals, about one-half inch in length, are white, with green "umbos" near the tip of each. The upper petal is purplish-blue; the lower ones are white with yellow hairs. It is similar to *Delphinium pavonaceum* growing further south in the valley which is larger, glandular, and grows in meadows rather than on rock. It blooms early in June.

87
Nuttall's larkspur
(*Delphinium nuttallii* Gray).
Buttercup Family (*Ranunculaceae*).
This species of larkspur is endemic to northwest Oregon, the Columbia Gorge, and southwest Washington. It is found in moist, open ground and near basaltic cliffs at lower elevations. This species may be threatened in Oregon but more information is needed to determine its status. It is being monitored in Washington.

Delphinium nuttallii grows sixteen to twenty-four inches tall and is densely puberulent. Its leaves are all stem leaves. The flower raceme is short with ten to fifteen flowers having dark blue sepals one-half inch in length, and a long, slender spur. The lower petals are blue and may be shallowly cleft. The upper petals are generally a lighter blue. It blooms in June and July.

88
Peacock delphinium
(*Delphinium pavonaceum* Ewan).
Buttercup Family (*Ranunculaceae*).
This delphinium is endemic to meadowland in the central Willamette Valley, and is found nowhere else in the world. As most of its natural habitat has been developed it is becoming rare and endangered throughout its range.

Peacock delphinium is similar to the White rock larkspur, but is larger, up to three feet tall, and is glandular-pubescent. It, too, has white sepals, three-quarters of an inch long with green "umbos" near the tips. The petals are much shorter than the sepals. The upper ones are dark bluish-purple; the lower are white, and have a tinge of purple at the base. It blooms May to June.

89
Purple toothwort
(*Dentaria gemmata* [Greene] How.).
Mustard Family (*Brassicaceae*).
This rare and threatened species is known from only a few sites in Curry, Josephine, and Jackson counties in Oregon, and in Del Norte and Siskiyou counties in California. It is known in California by an earlier name, *Cardamine gemmata* Greene, and is considered there to be endangered. It grows in wet places from an orange-yellow, egg-shaped rhizome.

The three to five basal leaves of *Dentaria gemmata* may be up to one and one-half inches long, are deep green, orbicular, shallowly lobed, and thick; the stem leaves are three to seven foliate. Several flowers form a short raceme. The petals are rose-purple to deep purple, one-half inch long on pedicels three-quarters of an inch long. The style is about one-quarter of an inch in length with a capitate stigma; the siliques are somewhat over an inch long. It blooms from April to June.

90 91

90
Oregon dicentra
(*Dicentra formosa* [Haw. in Andr.] Walp. *ssp. oregana* [Eastwood] Munz).
Bleeding Heart Family (*Fumariaceae*).
This rare white subspecies of the common bleeding heart is endemic to serpentine soil or rock in southwest Oregon and nearby California. It is easily confused with albino forms of the common species, creating some taxonomic confusion and disagreement.

The petals of *Dicentra formosa ssp. oregana* are yellowish-white with rose-colored tips. The long-stemmed, glaucous basal leaves are biternately compound and have a bluish tone. It blooms from April to June.

91
Firecracker flower
(*Dichelostemma ida-maia* [Wood] Greene).
Lily Family (*Liliaceae*).
In this case the botanical name is more confusing than the common name. Prior to 1987 it was called *Brodiaea ida-maia* (Wood) Greene in Oregon. In California (Abrams) it is referred to as *Brevoortia ida-maia* Wood, differentiated from *Brodiaea* and *Dichelostemma* by its long stipitate capsule. Although it grows in open woods, grassy hillsides, and roadsides from Douglas County, Oregon south through the Siskiyous into California where it is more common, it is seldom seen in Oregon.

The scape is tall, up to forty inches, the three linear leaves are much shorter. The stem is topped by an umbel of five to fifteen nodding flowers, tube-shaped, one to one and one-half inches long, brilliantly red with greenish-yellow lobes that are obtuse and often reflexed. There are three anthers alternate with three broad staminodia (sterile stamens). It blooms May to July.

Close-up of flowers

92
Dimeresia, also called Doublet
(*Dimeresia howellii* Gray).
Composite Family (*Asteraceae*).
This small rare plant of the high desert country of the Great Basin is found in southeastern Oregon from Baker to Lake County. It also grows in northwest Nevada, and in northeast California where it is considered to be endangered.

Dimeresia howellii is a mat forming annual, less than four inches in diameter, that grows in the dry sandy or gravelly soils. The stems are fleshy; the spatular-shaped leaves have entire margins, and a depressed central vein. Its base is beset with entangled, cobwebby hairs; the upper stems and leaves are somewhat glandular. The flower heads are all discoid, and form in a dense cluster in the center of the plant, each head having two to three tubular flowers about one-quarter of an inch long. The corolla tubes are purplish, the flaring lobes white to pink or yellowish. It blooms from May into August.

93
White shooting star
(*Dodecatheon dentatum* Hook.).
Primrose Family (*Primulaceae*).
This is the only true white shooting star in Oregon, although one occasionally sees a white *forma alba* in some of the other pink species. This rare species may be found at scattered sites on wet cliffs and moist ground in the Columbia River Gorge, and in the Clackamas River drainage. It ranges north to British Columbia and east to Idaho and Utah.

The white petals of *Dodecatheon dentatum* are one-half inch in length, strongly reflexed, and slightly yellowish with an undulated reddish-purple ring at the base. The stamens are generally a deep reddish-purple. The anthers which taper gradually to a minute two-toothed apex, form a tube around the style which extrudes slightly with a small, rounded stigma. There may be from one to five flowers on a stem. The leaves are ovate and sharply toothed and have long stems. The plant grows to about eight inches in height. It blooms in May and June.

94

Narcissus shooting star

(*Dodecatheon poeticum* Hend.).
Primrose Family (*Primulaceae*).
This relatively rare, regional endemic is found in the Columbia Gorge, and along the lower Deschutes River, in Wasco, Hood River, and Sherman counties in Oregon, and north to Yakima County in Washington. It prefers meadows and open woodlands that are moist in early spring.

Dodecatheon poeticum grows to six inches tall, is glandular-puberulent, and may have five to ten flowers on a single stem. Its reflexed petals, five in number, each about three-quarters of an inch long, are rose-purple, blending to white, then yellow near the base. The yellow is divided by a wavy purple line. The filament tube is dark purple. The lightly toothed leaves are ovate-lanceolate in shape, are puberulent on both surfaces, and are about four to five inches long. It blooms early, in March and April.

95

Ciliolate douglasia

(*Douglasia laevigata* Gray *var. ciliolata* Const.).
Primrose Family (*Primulacaea*).
This rare, low, rock-loving plant is usually found growing on the faces of cliffs. *Var. ciliolata* is found near the summit of Saddle Mountain in the north Oregon Coast Range, also in Hood River County, Oregon, and in Washington.

The leaves of *Douglasia laevigata var. ciliolata* are clustered, matted, and usually ciliolate, especially on the margins. The flowering stems are leafless, and covered with a whitish scale-like puberulence. The flower umbel has two to six deep rose-pink flowers that are about one-half inch long. Each flower is constricted where the tube flares into the five rotate, spreading lobes. The reddish-green calyx is about one-half as long as the flower, and has narrow sharply-pointed lobes. It blooms in May.

96
Smooth-leaved douglasia
(*Douglasia laevigata* Gray *var. laevigata*).
Primrose Family (*Primulaceae*).
Very similar to the ciliolate douglasia, this variety is found in the Columbia Gorge, southward in the Cascades to central Oregon, and north into Washington. It grows in subalpine and alpine zones on cliffs and steep rocky hillsides. More information is needed to determine its degree of rarity.

The leaves of *Douglasia laevigata var. laevigata* are thicker, but not as broad as those of *var. ciliolata*, and are not ciliated. Each flowering stem has from two to six flowers that are rose-pink, though generally paler than those of *var. ciliolata*, and the individual flowers have longer pedicels. It blooms from June to August.

97
White downingia
(*Downingia laeta* Greene).
Bluebell Family (*Campanulaceae*).
Also called the Great Basin downingia, this species of alkaline soil in California and Nevada is found in Oregon only in Lake and Harney counties.

Downingia laeta is a small plant, three to six inches tall which grows in shallow water and mud. Its stems are leafy, stout at the base, tapering decidedly to the inflorescence. The corolla is white or pale lavender, having markings of yellow and purple at the base of the three-lobed lower lip. The two lobes of the upper lip are slightly longer than those of the lower lip. The sepals are about equal to both. The stamens form a striped, greenish curved column, about three-quarters the length of the corolla. This species blooms in June.

98 99

98
Cascade downingia
(*Downingia yina* Applegate *var. yina*).
Bluebell Family (*Campanulaceae*).
This variety is found in bogs and wet meadows in the southern Oregon Cascades, and in Siskiyou County, California. It is rare and possibly endangered in Oregon, but more common in California.

Downingia yina var. yina reaches a height of three inches. Its leaves are linear-lanceolate, glabrous, and less than an inch long. The flowers, under one-half inch long, have a narrow tube which flares into two lips. The upper lip ends in two linear, erect lobes; the lower into three larger rounded, abruptly pointed lobes. The corolla is purple with a central yellow area surrounded by white on the lower lobes. Two yellow ridges extend from the lower lobe into the tube of the corolla. The stamen column does not extend beyond the tube. It blooms in July and August.

99
Golden alpine draba
(*Draba aureola* Wats.).
Mustard Family (*Brassicaceae*).
This rare mustard which grows on volcanic rock above timberline skips along the crests of the Cascades from Mt. Rainier south to the Three Sisters area of Oregon and into California as far as Lassen Peak. It is currently stable in Oregon and Washington and is considered to be endangered in California.

The branches of *Draba aureola* are up to eight inches long, decumbent, and very pubescent throughout, with white, branched to stellate hairs. The stems are very leafy up to the inflorescence. The flowers have four yellow petals. The broad, oblong, flattened, highly pubescent seed pods are about one-half inch long. It blooms in July and August.

100

Howell's whitlow-grass

(*Draba howellii* Wats.).

Mustard Family (*Brassicaceae*).

This rare endemic plant may be threatened throughout its range. It grows on high, rocky summits in the Siskiyou Mountains in southern Josephine County in Oregon, and in Humboldt, Shasta, Trinity and Siskiyou counties in California.

The flowering stems of *Draba howellii*, about two and one-half inches long, are capped with six to eight bright yellow flowers with petals one-third of an inch long. The style is longer than those of most yellow-flowered drabas. The puberulent seed pods are oblong, curved, pointed, and about one-quarter of an inch long. The leaves, which are in dense clumps, are less than one-quarter of an inch in length. They are broadly-spatulate with entire margins, and are somewhat shiny even though covered with fine, forked hairs on both surfaces. It blooms from June to August.

101

Cusick's draba

(*Draba sphaeroides* Pays. *var. cusickii* [Robins.] C. L. Hitch.).

Mustard Family (*Brassicaceae*).

This rare variety is found chiefly in the Steens Mountain area of southeast Oregon, and south into northern Nevada. It is limited in abundance but currently stable. It grows in rocky outcroppings at high elevations, about 6500 feet, on Steens Mountain.

The leaves of *Draba sphaeroides var. cusickii* are sessile, obovate, and covered with whitish forked hairs throughout. It has four yellow petals; its finely puberulent fruits are flattened siliques, which are oblong-lanceolate, and about one-quarter of an inch long. This perennial blooms in July and August.

102

Drummond's mountain avens

(*Dryas drummondii* Rich. ex Hook.).
Rose Family (*Rosaceae*).
This circumboreal species of Alaska and Canada reaches its southern limit in the Wallowa Mountains in Oregon, where it is rare but stable. It also occurs in Pend Oreille and Snohomish counties in Washington where it is referred to as sensitive. It grows on rocky slopes, generally above timberline.

Dryas drummondii is a prostrate shrub, about ten inches tall, with elliptical leaves, about one and one-half inches long on stems of equal length. The leaves have scalloped margins, and are shiny dark green above and white-tomentose beneath. The flowers, single on the top of a leafless stem, are nodding and have a calyx covered with black hairs. The eight to ten petals are bright yellow, flaring out from a center of numerous stamens that are attached to a saucer-shaped surface called a hypanthium. As it goes to seed the head becomes erect, twisted, and conical. The numerous styles elongate, and the flower head later becomes white and fluffy, with feather-like achenes (small, dry, one-seeded fruits). It blooms from June to August.

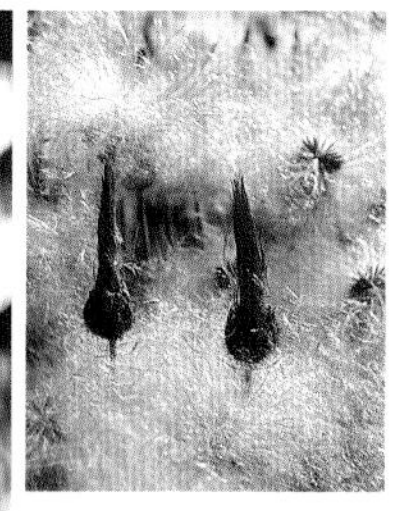

Two immature seedheads surrounded by ripened plumose achenes

103

White Mountain avens

(*Dryas octopetala* L. *var. hookeriana* [Juz.] Breit.).
Rose Family (*Rosaceae*).
This plant of the north reaches its southern limit in Oregon in the Wallowa and Blue mountains, where it is rare but stable. It is more common north into Washington and Canada, and east into Montana, Idaho, and Colorado. In Oregon it grows on rocky slopes and ledges at high elevations.

Dryas octopetala var. hookeriana is very similar to *Dryas drummondii*, differing by having eight white, spreading petals, and sepals without long, black hairs. The shiny, dark green leaves, with scalloped margins, are one-third narrower than those of *Dryas drummondii*. The styles, though less numerous, also become feather-like in seed. It blooms in July and August.

104
Sea-cliff stonecrop or Powdery dudleya
(*Dudleya farinosa* [Lindl.] Britt. & Rose).
Stone Crop Family (*Crassulaceae*).
This unusual and rare Oregon plant is found only on sea cliffs and the walls of sea stacks within the ocean spray. It grows in isolated populations from Bandon in Coos County southward into California. The specimen photographed was growing on the seaward side of a stack near the mouth of Pistol River, Oregon. It is rare and threatened in Oregon, but more common in California.

Dudleya farinosa ranges in size from about six to sixteen inches tall. The stems and leaves are fleshy, reddish in color, and partially covered with a whitish, waxy-like substance which gives it one of its common names. Its flowers are pale yellow and about one-half inch long. It blooms from May to September.

105
Phantom orchid or Snow orchid
(*Eburophyton austiniae* [Gray] Heller).
Orchid Family (*Orchidaceae*).
Although the phantom orchid, which ranges from the Olympic and Cascade mountains of Washington to southern California and east to Idaho, is no longer considered to be rare or threatened in Oregon, it is still a truly rewarding and exhilarating experience to find it in the wild. In Idaho it is still being monitored for threats to its stability.

Eburophyton austiniae is a mycotroph without chlorophyl, growing from the moist, cool ground in deep, chiefly coniferous woods, reaching a height of from eight to twenty inches. Each plant has from five to twenty flowers with sepals and petals about equal in length, one-half to three-quarters of an inch long. The entire plant is ghostly white except for a yellowish gland within. The lateral sepals are spreading, the upper sepal and petals are erect. It blooms from May to July.

106 107

106
Elmera
(*Elmera racemosa* [Wats.] Rydb. *var. puberulenta* C. L. Hitchc.).
Saxifrage Family (*Saxifragaceae*).
Originally known only in the state of Washington, this plant recently has been found on Three Fingered Jack, Cowhorn Mountain, and a few other peaks in Oregon, growing at high elevations in rocks and talus. It is limited in abundance but currently stable.

The stems of *Elmera racemosa var. puberulenta* are up to ten inches tall, and are strongly glandular-puberulent. The stems are topped with a loose raceme of small flowers with five broad greenish-yellow sepals, which are alternate with five small, narrow, white petals that are tipped with five to seven minute "fingers". The round leaves are toothed along the margins, and are smooth and glabrous on the top surface. The plants bloom in August.

Close-up of flowers

107
Black crowberry
(*Empetrum nigrum* L.).
This is our only species, and genus, of the Crowberry Family (*Empetraceae*) in Oregon.
A semi-shrub, it grows on moist rock and bluffs along the Oregon coast from Lane County south into California. It is also found on Mt. Rainier in Washington, and across the far north around the world, also in South America.

Empetrum nigrum is recognized by its prostrate, mat-forming stems which are several feet in length. It has dark green, linear, shiny evergreen leaves that are from one-quarter to one-third of an inch in length, have revolute margins, and a groove the length of the leaf on its under surface. The minute flowers grow singly in the leaf axils. Some are staminate; some pistillate; some are both. The flowers are globe-like, have three sepals, and may or may not have small purplish petals. The fruit is black and glaucous, and though used for food by the Eskimos, it is considered inedible or toxic here. The flowers bloom in June and July.

Ripened fruits

Female cones

108
Nevada ephedra
(*Ephedra nevadensis* Wats.).
Ephedra or Gnetum Family (*Ephedraceae*).
This strange plant has many common names, especially in the desert country south of Oregon. It is called Joint fir, as it is closely related to fir, and Mormon tea, as it makes a good tea which was used by early settlers. A Great Basin plant of Nevada, Utah, Arizona, and California, it is found at the northern limit of its range in southeast Harney County, Oregon growing on dry, gravelly slopes and flats.

Ephedra nevadensis is a three to six foot bush of glaucous, glabrous, grayish-green stems, with pairs of small scale-like, deciduous leaves that sheath the stem. In this species the stems are very divergent or spreading. The plants are dioecious, having male and female cones on separate plants, or monoecious, having both male and female cones on the same plant. They do not, however, have individual cones that are both staminate and ovulate. The cones "flower" in August, and later produce a pointed, round, sessile fruit, exserted slightly from the fruiting bracts.

Male flowers in cone

109
Green ephedra or Green mormon tea
(*Ephedra viridis* Coville).
Ephedra or Gnetum Family (*Ephedraceae*).
This ephedra, like the Nevada ephedra is found in Oregon, only in Harney County. Its range extends south into Nevada and California.

Green ephedra is very similar to Nevada ephedra, differing mainly in its manner of growth and color. It is bright green to bright yellowish-green, and instead of having divergent, spreading branches, they grow in an erect, clustered manner.

110
Broad-leaved willow-herb
(*Epilobium latifolium* L.).
Evening Primrose Family (*Onagraceae*).
Found in Oregon only in the Blue and Wallowa mountains, the range of this plant extends north into Washington, Canada, and Alaska, and east into Idaho, Montana, and Colorado. It is also native to Europe and Asia. Although it is more common in other parts of the world, it is apparently rare in Oregon.

Epilobium latifolium grows up to two feet tall on streambanks, river bars, and rocky slopes. The numerous leaves which are up to two and one-half inches long, are opposite, broad and thick, powdery green, and sometimes covered with fine, short hairs. The flowers, coming from leaf axils near the top of the stem have four pinkish-purple, broadly ovate petals about three-quarters of an inch long. It blooms from July to September.

111
Oregon willow-herb
(*Epilobium oreganum* Greene).
Evening Primrose Family (*Onagraceae*).
At one time this plant was thought to be a hybrid of the more common smooth willow-herb (*Epilobium glaberrimum* Barb.) which is smaller flowered. It is now considered to be a separate and distinct species. This rare and endangered plant is known only from Josephine County, Oregon, and three counties in northern California.

Growing in bogs at low elevations, *Epilobium oreganum* reaches eighteen to thirty inches in height. Its herbage may be glabrous or nearly so. The lance-shaped leaves are sessile, slightly toothed, from one to three inches long, and somewhat glaucous. The flowers are erect, with four pink to purplish colored petals that are slightly less than one-half inch long. The seed pods may be from one and one-half to two inches long, and have short pedicels. It blooms in June and July.

112
Rigid willow-herb
(*Epilobium rigidum* Hausskn.).
Evening Primrose Family (*Onagraceae*).
Also called Siskiyou Mountains willow-herb, this rare, threatened plant is endemic to serpentine soils in the Siskiyou Mountains of southwest Oregon and adjacent California.

The stems of *Epilobium rigidum* are up to sixteen inches long, somewhat decumbent, and sometimes forming mats. The leaves, one and one-half inches long, are numerous, crowded, thickened, generally entire, and glaucous. The flowers grow in a short terminal raceme. The calyx is purple, the petals are rose-purple, one-half to three-quarters of an inch long. There are eight stamens, four of them being about one-half the length of the others. It blooms in July and August.

113
Dwarf golden daisy
(*Erigeron chrysopsidis* Gray *var. brevifolius* Piper).
Composite Family (*Asteraceae*).
This variety, though quite abundant in its habitat, is a regional endemic limited to the alpine slopes of the Wallowa Mountains in northeastern Oregon.

Erigeron chrysopsidis var. brevifolius is a low growing perennial, about four inches tall. It is covered with a partially appressed pubescence. The leaves are narrow, linear, two and one-half inches long, and mostly basal. The flower heads are singular on the stem; the ray flowers are yellow, and less than one-half inch long. The disk flowers also are yellow, forming the central disk which is about one-half inch across. It blooms in July and August.

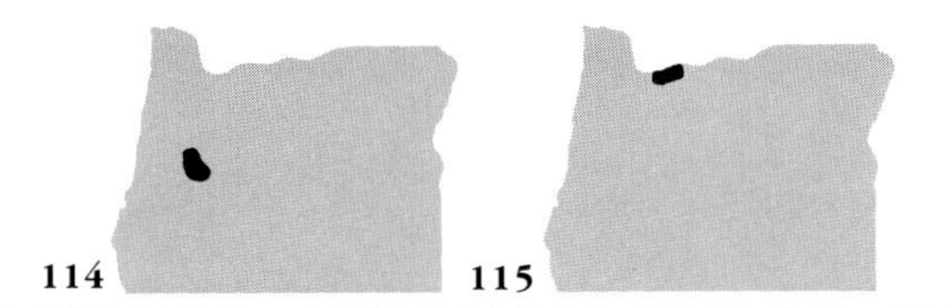

114
Willamette Valley daisy
(*Erigeron decumbens* Nutt. *ssp. decumbens*).
Composite Family (*Asteraceae*).
Once a very common plant of heavy soils on the native Willamette Valley prairies and grasslands, it nearly became extinct due to the destruction of its habitat through agriculture and land development. Last recorded in 1934, it was not rediscovered until 1980, when two sites were found. It is considered to be endangered throughout its range.

Erigeron decumbens ssp. decumbens is a perennial, growing from a taproot, six to twenty-eight inches tall. The leaves are linear and long, both basal and cauline; most of them are triple-nerved. The cauline leaves are only gradually reduced in size and extend nearly to the flowering heads. The heads may be one or many, and each has from twenty to fifty blue or lilac ray flowers that are about one-half inch long. The disk flowers are yellow; the disk is about one-half inch wide. It blooms from June into early July.

Close-up of flowers

115
Howell's erigeron
(*Erigeron howellii* Gray).
Composite Family (*Asteraceae*).
This narrow regional endemic is found only in rocky areas at the west end of the Columbia River Gorge in Multnomah County, Oregon, and Skamania County, Washington. It is described as limited in abundance in Oregon, and threatened in Washington.

Erigeron howellii is a perennial, up to twenty inches tall, glabrous except for some pubescence immediately below the flower. The lower leaves are stemmed; the upper ones clasping. All are rather broadly ovate. The flower heads are solitary, and have thirty to fifty white ray flowers about three-quarters inch long. The center of yellow disk flowers is about three-quarters inch across. It blooms from May to July.

116
Wandering daisy

(*Erigeron peregrinus* [Pursh] Greene *ssp. peregrinus var. peregrinus*).
Composite Family (*Asteraceae*).

This taxon is common in Alaska and British Columbia, but found in Oregon only in Clatsop and Tillamook counties in the north Coast Range. It is rare and threatened in Oregon but more common elsewhere.

Erigeron peregrinus ssp. peregrinus var. peregrinus grows up to twenty inches tall on a sturdy stem. Each flower head has forty to sixty ray flowers varying from nearly white to pink, sometimes lilac; each ray up to five-eighths inch long. These surround a broad area of yellow disk flowers. It blooms in June and July.

Close-up of flowers

117
Golden buckwheat

(*Eriogonum chrysops* Rydberg).
Buckwheat Family (*Polygonaceae*).

This plant, thought to be extinct, was rediscovered on a rocky tableland south of Harper in Malheur County in 1988. Said to be a species of Steens Mountain by its discoverer, William Cusick in 1901, it has been concluded that Mr. Cusick had extended the presently known definition of the Steens Range forty-five miles to the northeast where the plant was rediscovered in the Dry Creek area. This explains why the species had never again been seen on Steens Mountain. Small and endangered, it grows on arid, exposed rocky ridges and tablelands.

The stems of *Eriogonum chrysops* are leafless, and grow only one to two inches high. Its crowded basal leaves form small mats, are spatular in shape, taper to a stemless attachment, and are covered with woolly, white hairs on both sides. The flowers, one-eighth inch long, are yellow, and stem from a five-lobed involucre to form small, dense heads. They bloom in May and June.

118 119

118
Cusick's buckwheat
(*Eriogonum cusickii* M. E. Jones).
Buckwheat Family (*Polygonaceae*).
This rare species, threatened throughout its range, is known in only three small disjunct areas of dry volcanic tuff in southeastern Oregon—one west of Burns, one south of Christmas Valley, and one near Steens Mountain.

Eriogonum cusickii is a small perennial. Its foliage forms dense cushions of many tiny spatular-shaped leaves which are covered with whitish, woolly-like hair on both surfaces. They average one-half inch in length, and are attached by short, narrow stems. The very compact, yellow flowering heads are at the top of shiny, glabrous, reddish stems that are one to two and one-half inches tall. Its flowers bloom from late May into July.

119
James Canyon buckwheat
(*Eriogonum diclinum* Reveal).
Buckwheat Family (*Polygonaceae*).
This species is known from serpentine outcroppings on only a few high peaks and ridges in the Siskiyou Mountains of southern Oregon and northern California. The plants on the summit of Mt. Ashland, originally identified as *Eriogonum incanum* Torr. & Gray, have now been determined to be *Eriogonum diclinum*. It is rare and endangered in Oregon but more common in California.

Eriogonum diclinum is a low-growing, matted perennial. It is dioecious, having male and female flowers on separate plants. The separate plants appear almost as two different species. The male flowers are yellow, about one-eighth inch long; the female, reddish-yellow and larger, up to one-third inch long. The small leaves, grayish-woolly on both surfaces, are less than one inch in length; the stems of the inflorescence may be six to eight inches tall. It blooms from June to September.

Male flowers

Female flowers

120
Hausknecht's sulfur buckwheat

(*Eriogonum umbellatum* Torr. *var. hausknechtii* [Dammer] Jones).
Buckwheat Family (*Polygonaceae*).
This taxon, which according to the authority, James Reveal, occurs only on Cooper Spur on Mt. Hood, has also been identified from the Cascades of southern Washington, to Marys Peak in the Oregon Coast Range, to the summit of Steens Mountain in Harney County, Oregon. Confusing to botanists, it has been described as a "taxonomic mess". This sulphur-flower blooms in alpine areas where it is covered and uncovered by volcanic sands, as the wind blows.

The glabrous, stipe-like flowers of *Eriogonum umbellatum var. hausknechtii* are yellow with some reddish markings. The rest of the plant is quite woolly pubescent. The elliptic-shaped leaves are white-woolly beneath; the upper surface is a grayish-green. It blooms in July and August.

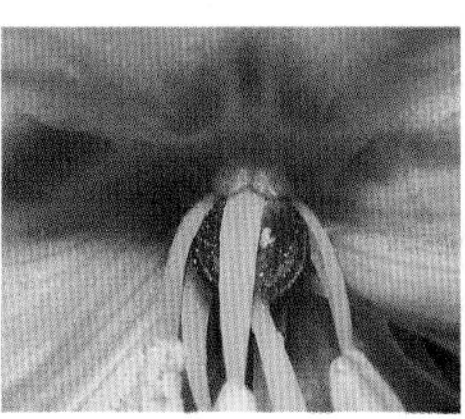

Saccate appendages at base of inner petals

121
Lemon-colored fawn-lily

(*Erythronium citrinum* Wats.).
Lily Family (*Liliaceae*).
Fawn-lilies also go by the common names of trout-lilies, adder's-tongues, dogtooth-violets, and lamb's-tongues. Those growing at higher elevations are called glacier lilies and avalanche lilies. This rare one is found only in the Siskiyou Mountains of Josephine and Curry counties in southwest Oregon where it grows in open woods and on brushy slopes.

The stem of *Erythronium citrinum* is six to eight inches tall. The mottled leaves, which are broadly lanced-shaped, and sharply tipped, are up to six inches in length. Its petals and sepals are one to one and one-half inches long, and are creamy-white to yellow in color, shading to a deep yellow or orange at the base of each. The style is clavate, and the stigma is very shallowly three-lobed. There are small appendages at the base of each of the three petals. The petals and sepals are recurved, and the tips tend to turn pink with age. It blooms in March and April.

122
Coast Range fawn-lily
(*Erythronium elegans* Hammond and Chamb.).
Lily Family (*Liliaceae*).
This recently described erythronium is also known as the Mt. Hebo fawn-lily because of the sizeable population on that mountain. It was once thought to be a hybrid between the pink coast fawn-lily (*Erythronium revolutum*), and avalanche lily (*Erythronium montanum*). This regional endemic is considered to be rare and threatened throughout its range.

Erythronium elegans grows eight to twelve inches tall, and has tepals up to two inches long. Its color varies from pink to nearly white, and there may be from one to ten flowers on a single stem. The tepals are lance-shaped, and slightly recurved at the tips. The stigma is deeply divided into three linear, curved lobes. The leaves are basal, close to the ground, and not mottled. It blooms from April to June, and may sometimes be seen blooming in a field of late snow.

123
Howell's adder's-tongue
(*Erythronium howellii* Wats.).
Lily Family (*Liliaceae*).
This rare lily is found in open woods primarily in the upper Illinois River basin, mostly in serpentine soil. It is considered threatened in Oregon and of limited distribution in California.

Erythronium howellii is very similar to the lemon-colored fawn-lily (*Erythronium citrinum*) having mottled leaves and whitish-yellow tepals with yellowish-orange areas at their base. There are three major differences: 1) *Erythronium howellii* has white anthers instead of yellow, 2) the stigma of *Erythronium howellii* is not distinctly divided into three lobes, 3) the petals, or inner tepals, do not have the saccate appendages at their base as do those of *Erythronium citrinum*. It blooms in April and May.

124
Klamath fawn-lily
(*Erythronium klamathense* Appleg.).
Lily Family (*Liliaceae*).
This rare regional endemic is found only in the open woods of the southern Cascades and Siskiyou Mountains of southwest Oregon and northern California.

Erythronium klamathense has pure white tepals with a brilliant yellow patch at the base of each. There are saccate appendages at the base of the inner petals. The tepals are three-quarters to one inch long. The stigma is unlobed, and both the stigma and the anthers are yellow. The leaves are broadly lance-shaped, deep green, without dark blotches. The plants normally grow three to eight inches tall, but may be found blooming as they come up from the ground. This species blooms in April and May.

125
Oregon fawn-lily
(*Erythronium oregonum* Appleg.).
Lily Family (*Liliaceae*).
Also known as the giant fawn-lily, and as the wild Easter lily, this species is found primarily at lower elevations in moist woods, on gravelly prairies, stony ridges, and damp hillsides at disjunct sites in the Willamette Valley and the coast ranges of Oregon and Washington. Its range extends north to Vancouver Island, and south into California.

Erythronium oregonum grows twelve to eighteen inches tall, and has creamy-white tepals that are yellowish-orange at their base. The flowers are pendent. The tepals may or may not be strongly recurved. The stigma of the pistil is deeply divided into three lobes. The stamens are broad at the base, and taper toward the anthers. The anthers are generally yellow, although there is a white-anthered variety found in the southern portion of the species range called *Erythronium oregonum var. leucandrum* Applg. The leaves are usually four to five inches long, green, with brownish to black mottling. As one goes further south in the range of this plant the mottling on the leaves becomes more defined and deeper black. This species blooms in April and May, sometimes into June.

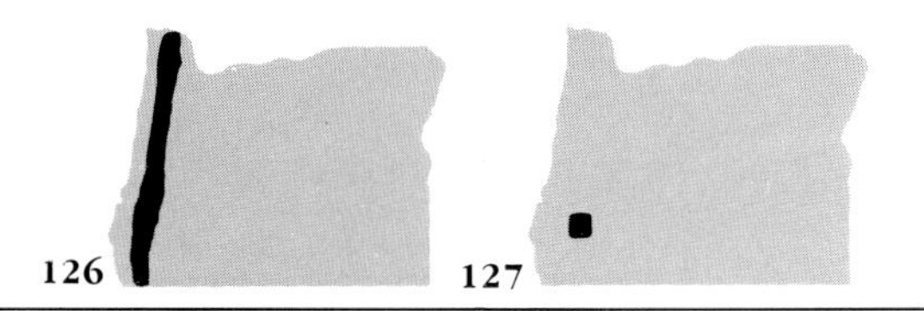

126
Coast trout-lily
(*Erythronium revolutum* J. E. Smith).
Lily Family (*Liliaceae*).
This species, rare and threatened by habitat disturbance, is found along the Oregon coast and in the Coast Range. Although it prefers shady conifer woods, and the margins of wet areas and streambanks, it has been found growing on the summits of Saddle Mountain and Neahkahnie Mountain. It is also known in Washington where it is considered a sensitive species.

Erythronium revolutum grows up to a foot tall and has tepals one and one-half inches long. The rose-pink tepals blend to yellow near their base, and each inner tepal has an elongated appendage at its base. The tepals are usually recurved. The stamens are broad and taper to the anther which is yellow. The stigma has three definite lobes. The leaves are about three and one-half inches long, and are mottled. It blooms in April and May.

127
Gold poppy
(*Eschscholtzia caespitosa* Benth.).
Poppy Family (*Papaveraceae*).
Though it is more common in California, it is known in Oregon from only two sites, southern Douglas County and northern Josephine County. It is considered to be endangered in Oregon, but more common elsewhere. It grows on dry, brushy slopes and flat areas, mostly along roadsides in Oregon.

Eschscholtzia caespitosa is a small annual poppy with flower stems that are from four to eight inches tall. Most of the leaves are basal. They are few in number, are several times divided into linear segments, and are much shorter than the flowering stems. The herbage of this plant is mostly glabrous and very glaucous, giving it a powdery, grayish-green appearance. The petals, usually four, are clear yellow, and less than an inch long. A differentiating characteristic of this poppy is the receptacle, or spreading rim of attachment for the petals; it is very narrow or almost non-existant. Most other species of the genus *Eschscholtzia* have broad, flaring receptacles. The seed capsule is elongated, linear, and about two to three inches long. This species blooms from March to early June.

128
Queen-of-the-forest
(*Filipendula occidentalis* [Wats.] How.).
Rose Family (*Rosaceae*).
This rare species, growing in rocks just above the high-water line along coastal rivers in Clatsop and Tillamook counties and on Onion Mountain, was severely reduced in numbers by the forest fires of the Tillamook Burn thirty to forty years ago, but has come back quite well in some areas.

Filipendula occidentalis is a tall-growing perennial herb, seven feet in height, with erect, simple stems. The leaves are pinnately divided into several remote leaflets that are quite small except for the terminal palmately-lobed leaflet which may be up to eight inches long, and as wide, and is divided into five to seven sharply toothed lobes. The panicle of white flowers is flat-topped; each of the five petals is one-quarter of an inch long, the five sepals are reflexed, and the stamens and pistils are numerous. It blooms late in May to August.

129
Umpqua swertia
(*Frasera umpquaensis* Peck & Appleg.).
Gentian Family (*Gentianaceae*).
Also known as *Swertia umpquaensis* (Peck & Appleg.) St. John, this plant was originally found in the vicinity of the Jackson-Douglas County line on the Rogue-Umpqua divide. It has now been found in Curry, Josephine, and Lane counties in Oregon, and south into California. It is limited in abundance but currently stable in Oregon; rare and endangered in California. It grows at elevations of 4500-6500 feet in conifer forests, in damp, shaded or sometimes open environments.

Frasera umpquaensis grows to four feet in height, and has a thick, stout, hollow stem. There are many long-stalked basal leaves that may be up to eight inches long, and several spatulate stem leaves in whorls of three that become smaller up the stem. Topping this tall stem is a dense, interrupted panicle of four-petalled flowers. The petals are yellowish-white to pale purplish-blue, veined with green, and each petal has a fringed gland near its base. The calyx is green, also four-parted, with narrow, linear segments that are longer than the petals. There are also some very short-stemmed flowers that bloom from the leaf axils lower on the stem. It blooms from June to August.

130 131

130
Diamond Lake fritillaria
(*Fritillaria adamantina* Peck).
Lily Family (*Liliaceae*).
Siddall, Chambers, and Wagner in their Interim Report of 1979 states that this species was thought to be extinct until it was found again near Diamond Lake in 1976. The same year it was found near the summit of Mt. Ashland in Jackson County. Some botanists are now saying that it is not a distinct species, but is instead a spontaneous hybrid of *Fritillaria atropurpurea* Nutt. It is included here for the possibility that it may again be considered a species in itself, as Peck differentiated it from *Fritillaria atropurpurea* by its stout stem, broad bulb, and many bulblets. This is not to encourage the digging of the bulb! It may be distinguished by its stout stem with narrow leaves, and its reddish, greenish, brownish perianth segments that are spotted with purple, and not recurved at the tips. It may grow up to two feet in height. It blooms from May to July.

131
Kamchatka fritillary
(*Fritillaria camschatcensis* [L.] Ker-Gawl.).
Lily Family (*Liliaceae*).
Also known as "indian rice" and "black lily", this species of meadows and moist open woods is found far north of here, from the Kamchatca Peninsula in Siberia to northern Washington. It has recently been found as a small disjunct population in the Bull Run watershed northwest of Mt. Hood in Oregon. It is rare and endangered in Oregon, but more common northward.

Fritillaria camschatcensis may reach sixteen inches in height, with flowers one and one-quarter inches long. The tepals are deep reddish-purple, streaked or mottled with yellow within, and greenish on the outside. The flowers are nodding. The leaves are lance-shaped with one to three whorls of three on the stem. It blooms here in June.

Capsule or seed pod

132
Falcate fritillary
(*Fritillaria falcata* [Jeps.] D. E. Beetle).
Lily Family (*Liliaceae*).
This species grows on rocky serpentine ridges in Stanislaus, Santa Clara, San Benito and Monterey counties in California and possibly on serpentine in southwest Oregon. A fritillary found recently on Chrome Ridge in the Oregon Siskiyou Mountains is thought by some to be *Fritillaria falcata*. If so, it is indeed rare and endangered in Oregon. It is also listed as endangered in California. The flower pictured was photographed on Chrome Ridge. Some of the leading botanists in the state, however, feel this flower is a form of *Fritillaria glauca*.

Fritillaria falcata has two to six fleshy alternate leaves at the base of the stem which are falcate, folded, and are from one and one-half inches to three and one-half inches long and less than one-half inch wide. The flowers with perianth segments nearly an inch long and one-quarter inch wide, are rusty-reddish-brown with yellow or green markings. The flowers, one to four, are erect, and terminal to a short stem. It blooms from March to May.

133
Gentner's fritillaria
(*Fritillaria gentneri* Gilkey).
Lily Family (*Liliaceae*).
This fritillary of Jackson and Josephine counties in southwest Oregon grows at lower elevations in dry, open fir and oak woodlands. It is rare and endangered throughout its range.

Fritillaria gentneri may reach two feet in height. The stout stem has several whorls of leaves in the middle portion. The flowers are nodding, are dull reddish-purple with yellow streaks within and without, and are up to one and one-half inches long. The petals may flare out at the tips, but are not recurved. It blooms in April and May.

134
Siskiyou fritillaria
(*Fritillaria glauca* Greene).
Lily Family (*Liliaceae*).
Found on dry, rocky serpentine slopes in Curry, Douglas, and Josephine counties in Oregon, this species is rare and endangered in Oregon but more common in California.

Fritillaria glauca grows only four to five inches tall. It has flowers up to an inch in length, which can be purple marked with yellow, or yellowish marked with purple and green. The alternate leaves are broadly-lanceolate, usually folded along the mid-line. The herbage is covered with a whitish "bloom". It blooms from April to June.

Yellowish form

Purplish form

135
Scarlet fritillary
(*Fritillaria recurva* Benth.).
Lily Family (*Liliaceae*).
This species once listed as threatened in Oregon due to habitat destruction and collecting is found in the woods from southern Douglas County through Josephine and Jackson counties into California.

Fritillaria recurva is a striking plant which grows to nearly two feet tall, and has flowers one and one-quarter inches long. The bright scarlet, nodding bells are spotted with yellow; their petals and sepals are curved back sharply at the tips. The stamens are nearly the length of the petals. The leaves are lance-shaped, one-quarter to three-quarters of an inch wide, two to three inches long, and are scattered or in whorls about the middle of the rather stout stem. It blooms from March into July.

136
Elegant gentian or Waldo gentian
(*Gentiana bisetaea* How.).
Gentian Family (*Gentianaceae*).
This gentian, first found near the deserted gold mining town of Waldo, is known from the upper Illinois and Chetco drainages in Oregon. It is threatened throughout its range. It grows in wet places, usually in *Darlingtonia* bogs. It is being suggested that *Gentiana bisetaea* is actually Mendocino gentian (*Gentiana setigera* Gray), and that what was previously thought to be *Gentiana setigera* is *Gentiana plurisetosa* C. Mason, ined. If this is true, the name *Gentiana bisetaea* will be dropped.

This gentian grows to fourteen inches in height. The flowers are one and one-half inches long, are deep blue inside, with a white center, sprinkled with green dots, and greenish on the outer surface. The funnelform flowers are erect and usually solitary at the top of the stem, or axillary to leaf-like, subtending bracts on the upper stem. The five petal lobes are separated by several long, thread-like, pointed appendages. The basal leaves are crowded; the stem leaves smaller, and generally paired. It blooms July to September.

137
Newberry's gentian
(*Gentiana newberryi* Gray).
Gentian Family (*Gentianaceae*).
This gentian of the high alpine meadows of the Cascade Mountains of Oregon from the Three Sisters area south into California is considered to be threatened in Oregon but more common in California. This species, however, may be the result of hybridization between the white form of *Gentiana newberryi* in southern Oregon and California and *Gentiana calycosa* Griseb. from farther north, consequently its taxonomic status remains uncertain.

Gentiana newberryi grows to five inches tall, often is prostrate. The flowers are large, up to one and one-half inches long, and are blue-violet inside and out, becoming lighter to nearly white toward the center on the inside. The petal lobes are dotted with green and terminate in a point. There are three appendages, that appear torn, between each of the five petal tips. It blooms August and September.

138
Moss gentian
(*Gentiana prostrata* Haenke ex Jacq.).
Gentian Family (*Gentianaceae*).
This very small gentian of high mountain meadows and alpine bogs, is native to a large area of the earth, from Alaska south through the Rocky Mountains to Colorado, west to Utah, Nevada, central Idaho, and the mountains of California, and around the world to Europe, Asia, and South America. Only recently found in Oregon near the summit of Steens Mountain, it is considered to be threatened in Oregon, endangered in California, but more common elsewhere.

Gentiana prostrata grows one to four inches tall; the stem has many pointed, oval, sheathing leaves. The flower is purplish-blue, about one-half to three-quarters of an inch in length, solitary and erect at the top of the stem. The stem may be erect or prostrate. The flower closes when shaded or touched. It blooms in July and August.

139
Western red avens
(*Geum triflorum* Pursh *var. campanulatum* [Greene] C. L. Hitchc.).
Rose Family (*Rosaceae*).
This plant is found only in grassy areas near the summit of Saddle Mountain in Oregon where it is endangered, and in the Olympic Mountains of Washington.

Geum triflorum is generally three to fifteen inches tall, and has three nodding flowers at the top of each stem. The calyx is reddish and consists of five broad, convergent, pointed lobes that are alternate with five outer, narrow, spreading calyx bracteoles. The petals, too, are convergent, are pinkish-yellow and are longer than the calyx lobes, differentiating this variety from a similar one, *Geum triflorum var. ciliolatum* [Pursh] Fassett, east of the Cascades. Later, seeds form with long feathery appendages which allow them to be carried away with the breeze. This has led to the use of another common name, "prairie smoke". It blooms in June.

140 141

140
Rough stickseed

(*Hackelia hispida* [Gray] Johnst.).
Borage Family (*Boraginaceae*).
This Washington species extends into Wallowa County of northeast Oregon and has been reported along Squaw Creek on the east side of the Cascades in Central Oregon. Two varieties of this species are listed in Washington, *disjuncta*, considered a sensitive plant, and *hispida*, believed to be extinct.

The stem of *Hackelia hispida* is simple or branched, up to twenty inches tall, and has long lance-shaped leaves with a stiff, short pubescence especially along the margins. The corolla is pale yellow to greenish-white in color and has short recurved lobes. The seeds are armed with glochidiate prickles (bristles having fine barbs at the tip) allowing them to stick to passers-by. It blooms in May and June.

141
One-flowered goldenweed

(*Haplopappus uniflorus* [Hook] Torr. & Gray *ssp. linearis* Keck).
Composite Family (*Asteraceae*).
This rare goldenweed is found at high altitudes in the Warner Mountains, on Steens Mountain and in the Ochoco Mountains of Oregon. This species grows on wet or dry open slopes and often in alkaline meadows.

Haplopappus uniflorus is distinguished from other goldenweed by having just one flower at the top of each stem, and by its long, narrow leaves that are three to five inches long, and one-eighth to one-quarter inch wide. Flower stems are four to twelve inches high, erect or lying on the ground. The plant is mostly glabrous. The ten to thirty ray flowers, each one less than one-half inch in length, appear as petals, and are bright yellow. The disk flowers are also yellow. It blooms from June to August.

142
Whitney's haplopappus
(*Haplopappus whitneyi* Gray *ssp. discoideus*
[J. T. Howell] Keck).
Composite Family (*Asteraceae*).
This California plant is known from only a few disjunct populations in the mountains of southwest Oregon. It grows in open forested areas on rocky slopes. It is a taxon that is threatened in Oregon but more common in California.

Haplopappus whitneyi ssp. discoideus is a perennial up to twenty inches tall, with several stems growing from a woody base. It is covered with a sticky, glandular pubescence throughout. The leaves are numerous, spiny-toothed, and are up to one and one-half inches long but gradually becoming smaller up the stem. The flower heads are terminal on the rather straw-colored stems. The involucral bracts under the flower head are about one-half inch long, and are layered like shingles over each other in three or four rows, with each bract flaring out and ending in a point. The flower head contains only yellow disk flowers with a pale reddish pappus; there are no petal-like ray flowers. The blooming time extends from July to September.

143
Purple large-flowered rush-lily
(*Hastingsia atropurpurea*).
Lily Family (*Liliaceae*).
Assuming it is a valid new species, this very recently described variety of *Hastingsia bracteosa*, known only in the vicinity of Woodcock Mountain in southwest Oregon, is rare and endangered throughout its range.

It is quite similar to *Hastingsia bracteosa*, differing chiefly by the deep purple color on the outer surfaces of its flower segments. It grows in a *Darlingtonia* bog, and reaches a height of twenty-eight inches. Its petals and sepals are one-half inch long, and it has stamens and pistils that are shorter than the tepals. It blooms in May and June.

144
Large-flowered rush-lily
(*Hastingsia bracteosa* Wats.).
Lily Family (*Liliaceae*).
Found only in the Siskiyou Mountains of southern Josephine and Jackson counties in Oregon, it grows in moist mountain meadows and along streambanks. Until the early 1980's this plant was known as *Schoenolirion bracteosa* (Wats.) Jepson. The name, *Schoenolirion*, was given the plant in 1922; it was previously named *Hastingsia* in 1885, thus this is a matter of returning to an older name.

Hastingsia bracteosa is similar to the Purple Rush Lily described above except for the floral color which is pure white. The size is about the same—the tepals are one-half inch long, and the stamens about two-thirds the length of the tepals. The bracts are narrow, tapering to a point, and are three-eighths of an inch long. The leaves, one to three, are ten to twenty inches long and approximately one-quarter inch wide. It blooms in May and June.

145
Northern sweet-broom
(*Hedysarum boreale* Nutt.).
Legume Family (*Fabaceae*).
This plant of the north occurs in Oregon only in the mountains of the northeast corner of the state where it is very limited. It is also found across Canada from the Yukon to Newfoundland and extends south in the Rocky Mountains to New Mexico and Arizona. It grows in gravel bars, along streams, on wooded hillsides, and in rock slides at elevations up to 8000 feet.

The flowering racemes may be five to fifty flowered with blossoms ranging from pink to reddish-purple. They are typically pea-shaped with banner, wings, and a keel, which is the longest, being nearly an inch in length. The seed pods usually have four to seven rounded segments, and are very plainly cross-corrugated. Its leaves are up to six inches long with seven to fifteen leaflets. They may be hairy on both sides or just underneath, and are about an inch long. This species blooms in July and August.

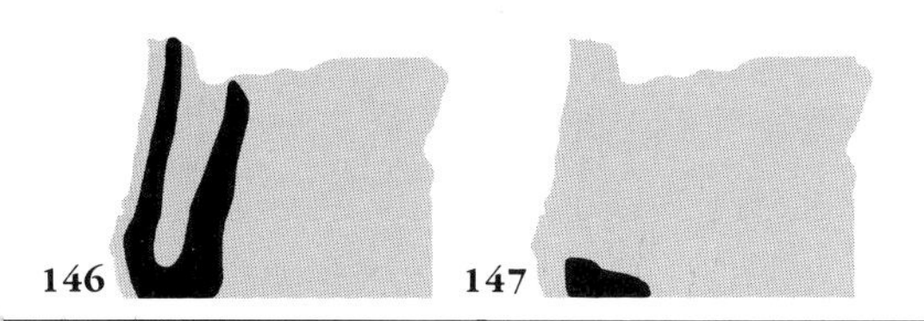

146
Gnome plant
(*Hemitomes congestum* Gray).
Heath Family (*Ericaceae*).
Some authors place this species in the Indian-pipe Family of plants (*Monotropaceae*) which includes all the closely related Heath Family saprophytic or mycotrophic plants. It is found in mature coniferous woods to moderate altitudes in the Cascades, along the coast, and in the Coast Range from the Olympic Mountains in Washington to central California.

Hemitomes congestum is a fleshy-white to brownish herb growing in clusters from a mass of fibrous roots. The congested stems grow to a height of two to six inches, are covered with very small scale-like, over-lapping, yellowish, mostly ovate leaves. The sepals may be two or four in number, linear and ciliate; the corollas are cylindric, one-half to three-quarters of an inch long, pinkish, hairy inside, and have from four to six petals which are unevenly cleft. The stamens, which are about three-quarters the length of the petals, are very villous, as is the pistil. They bloom from May to July.

Close-up of flowers

147
Bolander's hawkweed
(*Hieracium bolanderi* Gray).
Composite Family (*Asteraceae*).
This rare plant is endemic to the Siskiyou Mountains of southwest Oregon and adjacent California. Once very common in this area but now found in only a few scattered sites, it is threatened in Oregon, but apparently more stable in California. It grows on dry open or forested slopes.

Hieracium bolanderi is a perennial plant with a rosette of basal leaves which are hirsute on the margins and upper surfaces, and nearly glabrous beneath. The stems, which reach twelve inches in height, are nearly naked of leaves except for a few very small, linear bracts. The stems are quite glabrous with possibly a few short gland-tipped hairs, as are the involucral bracts beneath the flower head. The bracts are narrow, pointed, blackish tinged, and somewhat layered; the smaller ones vary in length. The yellow flowers are ligular. They close at night and open in the morning. It blooms in June and July.

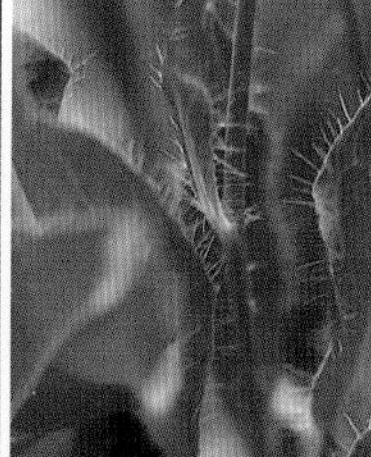

Long hairs on basal leaves

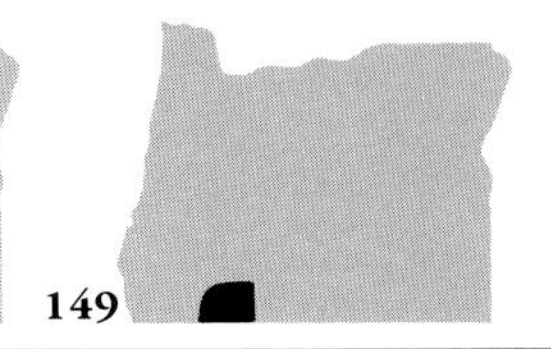

148
Long-bearded hawkweed
(*Hieracium longiberbe* How.).
Composite Family (*Asteraceae*).
This species, endemic to cliffs and rocky bluffs in the Columbia River Gorge in Multnomah and Hood River counties in Oregon and Skamania County, Washington, is rare throughout its range but currently stable.

Growing from one to two feet tall, this species is a perennial with upper stems that are nearly glabrous, but has leaves with long, rather bristly hairs almost one-half inch in length, which are particularly noticeable along the midrib and margins. The involucral bracts beneath the flower head also have the same kind of hairs stemming from a dark green, often black-based mid-rib. The basal and lower stem leaves wither early. The bright yellow flowers are ligular; when open they are about an inch across. It blooms in June and July.

149
Carrot-leaved horkelia
(*Horkelia daucifolia* [Greene] Rydb.).
Rose Family (*Rosaceae*).
This rare plant is found only from the southern part of Jackson County, Oregon, south to the Shasta Valley in California. It prefers dry open woods.

Horkelia daucifolia grows to twelve inches in height. The stems are reddish in color, hirsute with long white hairs, and are erect or bent over at the base. Most of the leaves are basal, very finely divided into as many as twenty-five leaflets which are each deeply cleft into four or five linear segments. The leaves are dull green, silky hairy, and up to four inches long. The inflorescence at the top of the stem is ball-shaped; it contains individual flowers with white to creamy, oval petals, that are attached by a slender base called a claw. The sepals are green, pointed, and somewhat shorter than the petals which are from one-quarter to one-third inch long. There are ten stamens with filaments that are broadly dilated at the base. It blooms from May to August.

150 151

150
Henderson's horkelia
(*Horkelia hendersonii* How.).
Rose Family (*Rosaceae*).
This very rare horkelia is endemic to the summits of a few granite peaks in southern Jackson County, where it is threatened throughout its restricted range. It apparently is not found in California.

Its stems are from five to eight inches high, with long, somewhat wavy hairs, and a few small leaves, which are subtended by mostly entire stipules. The predominately basal leaves are pinnate with leaflets that are palmately divided, and very hairy throughout, but not glandular. The flower is dense and hairy. The somewhat reddish, broadly lance-shaped sepals are about equal in length to the small white, linear-oblong petals, and the linear filiform bractlets that attach just below the sepals. It blooms from June to August.

151
Howellia
(*Howellia aquatilis* Gray).
Harebell Family (*Campanulaceae*).
Howellia is one of the rarest plants in Oregon. This rare and endangered plant is possibly extinct in this state. At present it is known from only two sites in Washington, in Clark County and Spokane County, and has recently been reported from several sites in Montana. It was collected in California in 1979, but efforts to relocate it have failed. It is presumed now to be extinct in that state.

Howellia aquatilis is an aquatic plant that is rooted in shallow ponds. It floats just under or near the water surface. The stems are lax and drooping, and up to thirty inches long. The narrow leaves are linear-tapering, mostly entire, and are about one to two inches long. The flowers are white to pale lavender, and are about an eighth of an inch across, but are not always present. When present, the corolla is bilabiate, the upper lip being very much smaller than the lower one which has five lobes. The early flowers are said to be cleistogamous, resulting in self-pollination, and are found in the axils of the ordinary leaves. Flowers later in the season form on special branches that have shorter leaves arranged about the leaf nodes of the stem in groups of three, as may be seen in the photograph. It flowers in May, and in some seasons when the ponds do not dry up, into August or September.

152

Alpine hulsea

(*Hulsea algida* Gray).

Composite Family (*Asteraceae*).

This alpine species is found in Oregon on a few high granitic peaks in the Wallowa Mountains. It is also found in Idaho, Montana, Wyoming, Nevada and the Sierra Nevada of California. It grows in gravel and sand.

Hulsea algida is a perennial. Each stem of this plant has a single flower head and but few leaves. The leaves, mostly basal, are linear, trough-like, and often toothed. They may be up to four inches long and three-eighths of an inch wide. The upper stem leaves are much smaller and are one to three white-nerved. The herbage is generally glandular-pubescent and green except for the bracts immediately under the flower head. They are white with dense, long hairs. The yellow flower heads are large, each with twenty-five to fifty ray flowers that are over one-half inch long. They are entire or slightly toothed at the apex, and bloom from June to August.

153

Ballhead waterleaf

(*Hydrophyllum capitatum* Dougl. ex Benth. *var. thompsonii* [Peck] Const.).

This narrow endemic of the scrub oak community at the east end of the Columbia Gorge grows from Wasco and Hood River counties, Oregon, to Yakima County, Washington. Within its restricted range it is relatively abundant.

The leaves of *Hydrophyllum capitatum var. thompsonii* are broad, oval-shaped with deep indentations dividing them into five or more lobes. The flowering stem, capped by a lavender, ball-shaped inflorescence, is taller than the leaves. The exserted stamens are twice the length of the flower petals. The stems, the leaves, and the inflorescence are covered with a dense, white pubescence. It blooms from March to June.

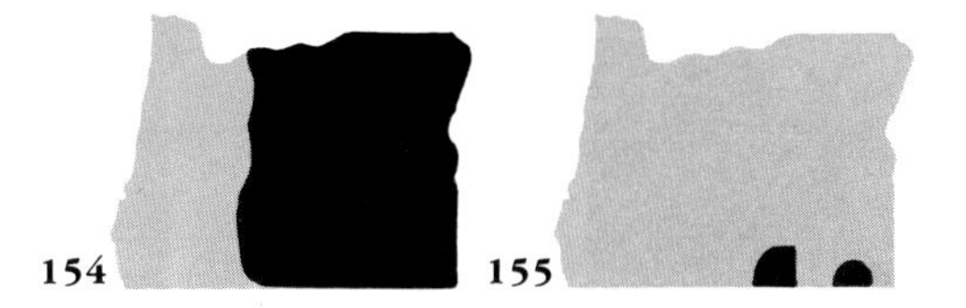

154
Columbia cut-leaf
(*Hymenopappus filifolius* Hook. *var. filifolius*).
Composite Family (*Asteraceae*).
A taprooted perennial plant that is quite widespread east of the Cascades in central Oregon, it is currently being reviewed for rarity status. It grows in dry areas at fairly low elevations.

Hymenopappus filifolius var. filifolius, often several stemmed, is mostly glabrous except at the base which is covered with very dense, white woolly hairs. The leaves are linear, once or twice pinnatifid. It grows from sixteen to thirty-six inches tall and is freely branched. The flower heads are discoid, having no ray flowers. The involucral bracts immediately supporting the tubular flowers are few in number, and occur in several series with slightly unequal lengths. The corollas are whitish or yellow. The stamens are minutely arrow-shaped. It blooms in May and June.

155
Cooper's goldflower
(*Hymenoxys cooperi* [Gray] Cockerell *var. canescens* [D. C. Eat.] Parker).
Composite Family (*Asteraceae*).
This species of dry, open areas in the Great Basin from southern Idaho, Nevada, eastern California to Arizona is found in Oregon in only a few isolated sites in Lake and Malheur counties. It is rare and endangered in Oregon, but more common elsewhere.

Hymenoxys cooperi var. canescens is a taprooted perennial with one to several stems up to twelve inches tall. Each stem branches into several shorter, terminal stems that are capped by solitary flower heads. There may be from three to thirty flower heads on a single plant. The stems are reddish, and rough with white hairs; the pinnate leaves with three to five divisions, are also rough and hairy. The flower heads are generally broader than tall. Both disk and ray flowers are yellow; the rays numbering from nine to thirteen, are quite widely separated, are about one-half inch long, and are tipped with three teeth. It blooms from May into the summer when the moisture disappears. When the plant dies it leaves a dry skeleton of itself standing until weather conditions break it down.

156 157

156
California globe-mallow
(*Iliamna latibracteata* Wiggins).
Mallow Family (*Malvaceae*).
This species is one of the more showy of Oregon's rare wildflowers, resembling closely the hollyhock in the same plant family. It is sometimes referred to as a wild hollyhock. It grows in the coastal ranges in southwest Oregon in Coos, Curry, Douglas, and Josephine counties, and south into Humboldt County, California. It is very rare and threatened in Oregon, uncommon, and being watched in California.

Iliamna latibracteata is distinguished by its rich rose-colored flowers, supported by bracts that are broader than those of similar species, and by its leaves which are green above, and covered with whitish hairs beneath. The plants are from twenty to fifty inches tall; the flowers have petals an inch long. It blooms in June and July.

157
Clackamas iris
(*Iris tenuis* Wats.).
Iris Family (*Iridaceae*).
A small iris, with stems less than twelve inches tall, it grows in the western Cascades of eastern Clackamas County, Oregon. It may be found on moist wooded hillsides and along a few streams in the Clackamas River drainage. It was once proposed for endangered status, but clear-cutting and powerline right-of-ways have opened up new habitats and allowed it to spread. However, it is still rare but not threatened.

Iris tenuis has slender stems, often with two flowers on a stem. It has broad, thin, soft leaves that may be taller than the flower stems. The flowers are pure white, delicately lined and shaded with yellow on its broad spatular-shaped sepals. The sepals are about one and one-half inches long, they terminate in a perianth tube that is less than one-quarter of an inch in length. The petals are narrow, erect and shorter than the sepals. It blooms in May and early June.

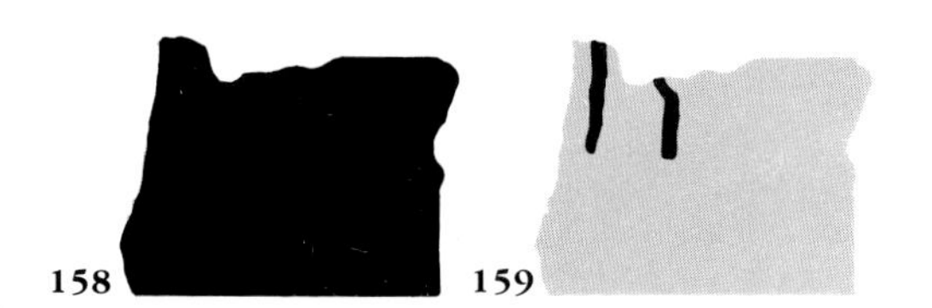

158
Howell's quillwort

(*Isoetes howellii* Engelm.).
Quillwort Family (*Isoetaceae*).
Quillworts are closely related to ferns and horsetails and are found growing in pools that dry up in the season. There are four species of quillworts in Oregon, all are widely distributed, but rarely seen, and very difficult to identify. They are all considered rare, and possibly threatened. *Isoetes howellii* is found from Washington to California and east to Montana.

The leaves of *Isoetes howellii*, ten to thirty in number, arise from a corm-like stock which is divided into two lobes. They are green, cylindrical, and divided into four longitudinal air spaces having several transverse partitions. They are erect, or nearly so, and are about five to eight inches long. This plant has a sporangium that produces spores, the reproductive bodies of fern-related plants. The sporangium is only partially covered by a velum, a fold in the inner side of the leaf base. The large spores are white and have short tubercles.

Closely related and having the same rarity status is Nuttall's quillwort (*Isoetes nuttallii* A. Braun ex Engelm.). It is limited to the Columbia River Gorge and west of the Cascades. It differs from *Isoetes howellii* by having a three-lobed stock at the base of the plant, and the sporangium is completely covered by a velum.

Howell's quillwort

Nuttall's quillwort

Two-lobed corm of Howell's quillwort

159
Hall's isopyrum

(*Isopyrum hallii* Gray).
Buttercup Family (*Ranunculaceae*).
This showy flower grows at scattered sites in damp woods, along streambanks, and on wet rocky walls in the foothills of the Cascade and Coast Range mountains of northern Oregon and southern Washington. Once considered threatened in both Oregon and Washington, they are now believed to be more abundant, though still rare.

Isopyrum hallii has flowers that have five to nine white sepals, no petals, and numerous stamens. The filaments are clavate, thin at the base and broadening toward the anther. The plants grow from twelve to thirty inches tall, are predominately glabrous, and have long-stemmed, deep green leaves that are deeply divided into three lobes. The blooming period is in June and July.

160
Bailey's ivesia
(*Ivesia baileyi* S. Wats., also known as *Horkelia baileyi* [Wats.] Greene from about 1890 to 1970).
Rose Family (*Rosaceae*).
This small rare plant grows at high elevations in the mountains of Harney County, Oregon and south into Nevada. It is usually found on the face of cliffs.

Ivesia baileyi has pinnately compound leaves that are three to four inches long. Each leaflet is deeply indented into several lobes. The flowers are small and saucer-shaped, with five broad, green sepals, and five narrowly-spatulate, white or yellowish petals. The sepals and petals are alternate with each other, and are nearly equal in length, about one-eighth of an inch. There is also a broad, very conspicuous maroon-colored, five-lobed disk (hypanthium). The five stamens are exserted peripherally through the notches of the disk.
It blooms in July and August.

161
Grimy ivesia
(*Ivesia rhypara* Ertter & Reveal).
Rose Family (*Rosaceae*).
The common name given this plant honors one of its discoverers, Jim Grimes. Discovered very recently, the original publication of its description was in 1977. One of the truly rare plants of the world, this species is endangered throughout its range. Its habitat is limited to a very specific loose volcanic ash at an elevation of about 4500 feet in dry eastern Malheur County, Oregon, and in northern Nevada.

Ivesia rhypara is a low, spreading, perennial with stems up to six inches long. Its extremely villous leaves are pinnate with as many as fifteen overlapping pairs of leaflets which are further divided into several segments. The flowers form at the end of the stem, the terminal one blooming first. The five green sepals are woolly, pointed, and alternate to the five minute, white, narrowly-spatulate petals, and the five linear, woolly bractlets that seem to cover the seams of the calyx while the flower is in bud. The center of the flower is made up of a shiny, honey-colored hypanthium with five stamens exserting from its margins, opposite the petals, and a thread-like pistil arising from its center. It would appear that ants are attracted to this species. It blooms from May to October.

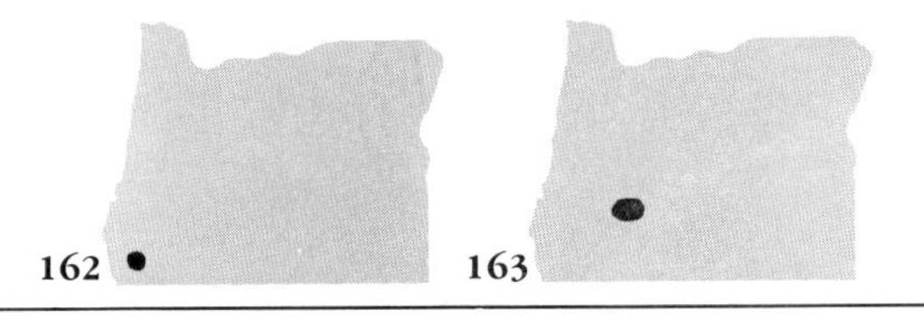

162
Kalmiopsis
(*Kalmiopsis leachiana* [Hend.] Rehder *var. leachiana*).
Heath Family (*Ericaceae*).
This very special plant is found only in the Kalmiopsis Wilderness Area in the Siskiyou Mountains of southwest Oregon. It was first discovered in May 1931, by Lilla Leach of Portland. The wilderness area later was named for it. This plant is unique in that it is believed to predate the ice age, and consequently is one of the oldest members of the Heath Family. It is quite similar to the genus *Rhododendron*, and was originally named *Rhododendron leachianum* Henderson, and at another time was named *Rhodothamnus leachianus* Copel, before the present taxonomic name was finally accepted. Although rare, it has not been listed as threatened or endangered, possibly because of its inaccessibility.

Kalmiopsis leachiana var. leachiana is a shrub growing to about ten inches tall. It has numerous, thick, deep green, elliptically-shaped leaves that are one-half inch long. The leaves are glandular-dotted beneath. The flowers are numerous and showy. The petals are deep rose, about one-half inch long; the bright red sepals are about one-third the length of the petals. The pistil and the exserted stamens also are bright red. It blooms from about Memorial Day through June.

Manner of growth

163
North Umpqua kalmiopsis
(*Kalmiopsis leachiana* [Hend.] Rehd. *var. nov.*).
Heath Family (*Ericaceae*).
Discovered in recent years on rocks high above the North Umpqua River in Douglas County, Oregon, this variety is still unnamed. It has been referred to by some as *Kalmiopsis lepiniac* but this name has not become official. It is very similar to *Kalmiopsis leachiana var. leachiana* and was once thought to be the same species. Though still officially unnamed this plant is currently on a review list in Oregon, as more information is needed to determine its true status. Its range is very limited.

The unnamed taxon is less shrub-like, more mat-forming, and vine-like than *Kalmiopsis leachiana var. leachiana*. Its long, woody stems have been traced deeply into the cracks of the hard rock walls on which it grows. It reaches only a few inches in height above the rock surface. The leaves of the Umpqua variety appear to be more numerous, slightly thicker, darker green, and lightly glandular on the upper as well as the lower surface. The flowers of the Umpqua variety may have a slightly different shade of pink, with the membranous margins of the corolla more deeply cleft. The stamens are about twice as long. It blooms from May to June.

164
Large-flowered goldfields
(*Lasthenia macrantha* [Gray] Greene *ssp. prisca* Ornduff; also known by the genus name *Baeria*.)
Composite Family (*Asteraceae*).
This rare daisy-like flower grows in a few isolated populations in Curry County on seaward slopes, rocky cliffs, and sandy areas above the beach. It is rare throughout its range, and its numbers seem to vary with the year; some years there are very few to be found.

Lasthenia macrantha ssp. prisca is a very showy plant about sixteen inches tall. The flower heads are singular on each stem, are about an inch in diameter, and usually have about twelve bright yellow ray flowers. The disk flowers are yellow, also. The leaves are linear, untoothed, and generally villous. It flowers in June and July.

165
Seaside goldfields or Hairy lasthenia
(*Lasthenia minor* [DC.] Ornduff *ssp. maritima* Gray Ornduff, or *Baeria maritima* Gray).
Composite Family (*Asteraceae*).
Once known from only a very few sites on rocky sea cliffs along the Pacific Coast from Vancouver Island to California, this maritime plant has now also been found on the "bird rocks" off shore where gulls and other sea birds nest and is no longer considered threatened.

Lasthenia minor ssp. maritima is a low, spreading herb with linear or oblong leaves, without teeth or slightly lobed on either side. The herbage varies from being sparsely hairy to glabrate. The flower heads are small, normally very short stemmed. The small (less than one-eighth inch long), yellow ray flowers are few, varying in number from three to eight, and are sometimes toothed at the apex. The disk is about three-eighths of an inch across and consists of numerous yellow flowers.It blooms from June to July.

166
Del Norte pea
(*Lathyrus delnorticus* C. L. Hitchc.).
Pea Family (*Fabaceae*).
This rare plant is an endemic to the Siskiyou Mountains in southern Curry and Josephine counties in Oregon and Del Norte County, California. It grows on serpentine soil, mostly in moist thickets and in wooded areas. This pea is considered to be threatened in Oregon, and limited in its distribution in California.

Ranging from eight to thirty inches tall, its stems are sometimes broadly winged, climbing by tendrils at the leaf tips, or trailing. Its leaves have from ten to sixteen narrowly elliptic leaflets not exactly paired with each other. The inflorescence is a loose raceme of eight to twenty flowers, each over one-half inch long, with a whitish to orchid banner that is lined with purple; the other petals are white. The lower flowers wither first, turning an amber color in age. The pods are about one and one-half inches long. It blooms in June and July.

167
Columbia bladder-pod
(*Lesquerella douglasii* Wats.).
Mustard Family (*Brassicaceae*).
This is a species of the sandy shores of the Columbia River, east of The Dalles, north into Washington and British Columbia. More information is needed to determine its degree of rarity.

A perennial plant, its erect or spreading stems may be from four to eighteen inches long, and are covered with silvery stellate hairs. The basal leaves are broad, some of them with deep, wavy margins that taper to a slender stem. They are densely stellate, and about two to three inches long including the stem. The upper stem leaves are narrow and linear with entire margins. The four petals are bright yellow, less than one-half inch long. The fruits are globose pods which are also covered with short star-shaped hairs. They are less than half the length of the petals, and are much shorter than the stems to which they are attached. It may be found in bloom from May to July.

168
King's bladderpod

(*Lesquerella kingii* Wats. *ssp. diversifolia* [Greene] Rollins & Shaw; has also been known as *Lesquerella sherwoodii* and *Lesquerella occidentalis var. diversifolia*).
Mustard Family (*Brassicaceae*).
It is found on talus and rocky slopes in the high Wallowa Mountains of northeast Oregon. Although locally abundant, its known sites are quite limited in number.

Lesquerella kingii ssp. diversifolia is characterized by its lax, prostrate stems, four to sixteen inches long, and its siliques which are flattened contrary to the partition. It is densely covered by stellate hairs giving its surface a whitish appearance. The basal leaves are nearly round, attached by long stems, the upper leaves much smaller and spatulate. The flowers are yellow, slightly over one-quarter inch long. It blooms in July and August.

169
Western leucothoe

(*Leucothoe davisiae* Torr. ex Gray).
Heath Family (*Ericaceae*).
This rare shrub is found in damp areas from Curry County, Oregon, south into California.

Leucothoe davisiae grows twenty to sixty inches tall; its one to three inch leaves are elliptical, stiff, glabrous, entire or finely toothed. The inflorescence is a tight raceme of twenty to thirty white, globose flowers at the top of an erect stem. There are ten stamens enclosed within the corolla. It blooms in June and July.

170 171

170
Columbia lewisia
(*Lewisia columbiana* [How.] Robins. *var. columbiana*).
Purslane Family (*Portulacaceae*).
This plant, found in Oregon in the Columbia River Gorge in Hood River and Multnomah counties, extends north in Washington to the Okanogan. It has also been reported on three mountains in the southeastern part of Douglas County. It is rare and threatened in Oregon, but more common in Washington. A similar variety, *wallowensis* C.L. Hitchc., can be found in the Wallowa Mountains of Oregon, and in Idaho, where it is also rare.

Lewisia columbiana var. columbiana is a perennial with several flowering stems ranging from four to ten inches tall. The many shiny, dark green basal leaves are up to two inches long, and are narrowly oblong to almost linear or spatulate. The stems are topped with a panicle of several white flowers. The panicle is subtended by bracts that are fringed with reddish glandular tips. Each flower has two sepals which are also fringed with reddish glands. There are seven to ten petals that are nearly one-half inch long, each with rose-colored veins. They are broadest and fringed at the tips. It blooms from May to July.

Close-up of flower

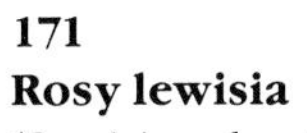

171
Rosy lewisia
(*Lewisia columbiana* [How.] Robins. *var. rupicola* [Engl.] C. L. Hitchc.).
Purslane Family (*Portulacaceae*).
Also called English's Lewisia, this rare plant is found in Oregon only on the summits of a few peaks in the Coast Range in Clatsop County (Saddle Mountain, Onion Peak, and Sugarloaf) and on Blue Lake ridge in Tillamook County. It is also found in Washington in the Olympic Mountains and near Mt. Rainier.

Although closely related to the previous variety, *columbiana*, it is considerably different in appearance. The succulent, angular-spatulate leaves, one and one-half inches long, often tipped with red, form perfectly geometrical rosettes. Arising from between the rosettes are red stems, five to eight inches tall, supporting an open inflorescence of ten to twenty-five flowers. The petals are rich rose-pink veined with deep red, oblong in shape and fringed at the tip. The subtending bracts and the two sepals of each flower are fringed with reddish glands. There are six stamens with arched and flaring filaments that are topped with bright red anthers. The pistil is also reddish; its stigma is deeply divided into three. It blooms in June and July.

Basal-leaf rosette

172
Howell's lewisia
(*Lewisia cotyledon* [Wats.] Robins. *ssp. howellii* [Wats.] Jepson).
Purslane family (*Portulacaceae*).
This striking subspecies, one of three which occur in Oregon, is found only in southwestern Oregon and adjacent California. It is rare throughout its range. It grows in shallow or rocky soil in oak woodlands at elevations of 500 to 1200 feet.

Lewisia colyledon ssp. howellii grows to twelve inches tall, its seven to nine petals are about three-quarters inch long. Each petal is closely veined with up to a dozen deep rose-colored lines that become further apart as they extend toward the rounded tip of the petal. Surrounding them is a border of pure, translucent white. Distinguishing this subspecies is the leaf. It has wavy margins. The leaves of the other subspecies have straight margins. The leaves are numerous and basal, fleshy, two to three inches long. The leaves on the flowering stems have been reduced to small bracts. The two sepals on each flower are glandular-tipped. It blooms in April and May.

173
Lee's lewisia
(*Lewisia leana* [Porter] Robins.).
Purslane Family (*Portulacaceae*).
This species of high elevation serpentine ridges in California reaches the northern limit of its range in southwest Oregon where it is both rare and threatened.

Lewisia leana is distinguished by its linear, cylindrical base leaves that are one to two and one-half inches long, and are covered with a whitish "bloom". There are from one to four leafless flowering stems, each many branched and ending in a loose, many-flowered inflorescence. Each flower has six to eight pinkish to rose-red petals that are striped with deeper red veins. The two sepals are tipped with dark glands. The petals are about one-quarter inch long. It blooms from late May to August.

174 175

174
Opposite-leaved lewisia

(*Lewisia oppositifolia* [Wats.] Robins.).
Purslane Family (*Portulacaceae*).
Found at elevations of 1000 to 4400 feet in serpentine rock or soil in places that are moist in spring, this species is found only in the Siskiyou Mountains of southwest Oregon and northern California. Though not currently threatened, it is limited in abundance and in its range.

Lewisia oppositifolia grows from two to ten inches tall. It has several narrowly lance-shaped basal leaves that are two to four inches long, and one to three pairs of opposite leaves on the flowering stem, which are similar to the basal leaves but much smaller. The flowers are few and have ten white to pinkish petals that are one-half to three-quarters inch long. There are two glandularly fringed sepals. There are many stamens; the stigma is deeply divided. It normally blooms from March to May.

Close-up of flowers

175
Bolander's lily

(*Lilium bolanderi* Wats.).
Lily Family (*Liliaceae*).
This very attractive lily, found only on rocky serpentine slopes in the Siskiyou Mountains of Josephine and Curry counties, Oregon and Del Norte County, California is rare and threatened throughout its range by collectors.

Lilium bolanderi grows twelve to thirty inches tall, has three to four whorls of rather wavy leaves with a few small ones between whorls. The nodding flowers, two to seven per flowering head, have broadly lance-shaped tepals about one and one-quarter inches long. They are deep wine red in color, turning yellow toward the center, with maroon colored dots extending to the tip. The tepals are only slightly spreading and very little recurved. The stamens are considerably shorter than the tepals, and have deep red anthers. It blooms in June and July.

176
Kellogg's lily
(*Lilium kelloggii* Purdy).
Lily Family (*Liliaceae*).
This species of lily is now believed to be possibly extinct in Oregon. It is apparently more common in California. It grows on a sandstone/sedimentary type of soil in dry wooded areas.

Lilium kelloggii is quite similar to *Lilium rubescens*, but has smaller tepals that are recurved to the base. It is pink with purple dots and has a touch of yellow along the midline of each tepal. It grows from three to five feet tall and has several whorls of long, narrow, lanceolate leaves, with as many as sixteen in a whorl. It is very fragrant, and it blooms in June.

177
Western lily
(*Lilium occidentale* Purdy).
Lily Family (*Liliaceae*).
One of the rarest plants in Oregon, this extremely rare lily which grows only on the periphery of bogs near the ocean from Coos County, Oregon to Humboldt County, California is endangered by collectors and by the destruction of its habitat.

Lilium occidentale grows up to five feet tall, and has as many as ten nodding flowers per stem. They are crimson red shading to yellow and green at the base. The yellow and green areas are dotted with purple. The tepals, two inches long, are recurved only halfway. The deep red anthers, one half inch long, closely surround the pistil. It has no fragrance. Its leaves generally are single along the stem except for one whorl near the middle. It blooms from late June through July.

178
Alpine lily
(*Lilium parvum* Kell.).
Lily Family (*Liliaceae*).
Also referred to as small tiger lily. This species is believed at one time to have existed in Curry, Josephine and Klamath counties in southwest Oregon. Apparently it has been some time since it was last reported in this state. Many now believe it to be extinct in Oregon. On the premise that it may still be found somewhere in Oregon, it is here included. It grows along mountain streams and in wet thickets at high elevations. It is apparently more common in the Sierra Nevada Mountains further south in California.

Lilium parvum grows to about five feet in height; its stem supports mostly scattered leaves, or a few partial whorls. The flowers, one and one-half to two inches long, are normally orange-yellow going to a dark red at the outer edges, spotted with maroon. The photograph shows a pink variety called *Lilium parvum var. rubellum*. It blooms in June and July.

179
Lilac lily or Chaparral lily
(*Lilium rubescens* Wats.).
Lily Family (*Liliaceae*).
This California lily was once known from a few sites in Josephine and Curry counties in Oregon. It is now considered possibly extinct in this state though presently stable in California. It grows on dry wooded slopes.

Lilium rubescens is a large lily, sometimes reaching seven feet in height. The long, stout stem supports several whorls of six to twelve leaves, four to five inches long, that are glaucous beneath. The two to eight flowers atop the stem are erect, pinkish white dotted with purple, gradually darkening to a rose-purple with age. The tepals are one and one-half to two and one-half inches long and reflexed for about one-third to one-half their length. The anthers are yellowish, about one-quarter of an inch long. The branching flowering stems ascend nearly parallel to the main stem. It blooms in June and July.

180
Vollmer's tiger lily
(*Lilium vollmeri* Eastw.).
Lily Family (*Liliaceae*).
This rarely seen lily which grows in hillside bogs, often with *Darlingtonia*, is found chiefly in Josephine and Curry Counties in Oregon and in adjacent northern California. It may grow in serpentine soil but is not restricted to it. It is threatened throughout its range chiefly by collectors. What used to be called leopard lilies (*Lilium pardalinum* Kellogg) in Oregon are now believed by some to be *Lilium vollmeri*.

Lilium vollmeri grows about three feet high. The leaves are scattered and erect on the stem, though some may be whorled. The one to several flowers are nodding; the recurved tepals, about two and one-half to three inches long, are dark red on the outer half, and yellow to orange with maroon spots near the center. The orange-brown anthers do not flare greatly from the pistil. It blooms from late June to August.

181
Wiggin's lily
(*Lilium wigginsii* Beane and Voll.).
Lily Family (*Liliaceae*).
This rare regional endemic, known in Jackson and Josephine counties in Oregon, and Humboldt and Del Norte counties in northern California, grows at fairly high elevations along stream banks and wet seeps and meadows.

Lilium wigginsii is a yellow lily with purple spots on its perianth segments which curve back from the base. It grows to about five feet tall. Its petals and sepals are two to three inches long. Its long, lance-shaped leaves, five to nine inches, are spreading and usually scattered along the stem, but may be in whorls of three or four leaves. Its blooming period is from June through July.

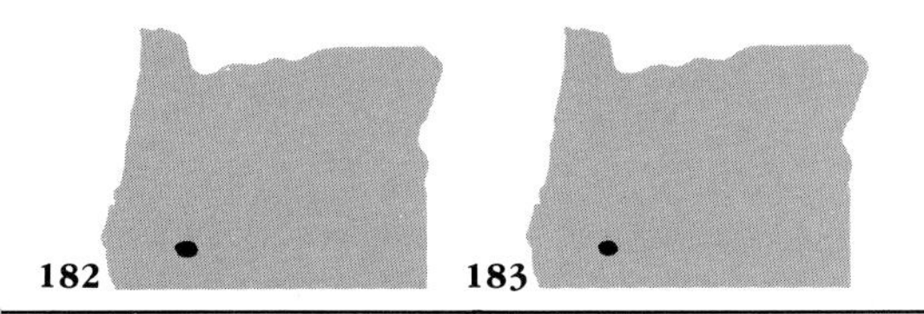

182
Big-flowered woolly meadow-foam
(*Limnanthes floccosa* How. *ssp. grandiflora* Arroyo).
Meadow-foam Family (*Limnanthaceae*).
This rare local subspecies of meadow-foam is known only from the "Agate Desert" in Jackson County, Oregon, where it grows in vernal pools that dry up later in the summer. It is endangered throughout its very restricted range.

Limnanthus floccosa ssp. grandiflora is a low annual, about six inches tall. The flowers are white, lined with green veins on the inside. The flower is quite large for the size of the plant. The sepals and the petals are about the same length, three-eighths of an inch. The stamens are about one-fourth inch long. The calyx is very pubescent inside and out, the narrowed base of the petals have two rows of woolly pubescent hairs. The stems and leaves are sparsely pubescent differentiating it from the *ssp. floccosa*. It blooms in April and May.

183
Dwarf meadow-foam
(*Limnanthes floccosa* How. *ssp. pumila* [Howell] Arroyo).
Meadow-foam Family (*Limnanthaceae*).
This is a very local subspecies found only on the top of two flat-topped basalt tablelands in Jackson County, Oregon, where it grows at the edges of vernal pools that dry up by mid-summer. It is considered to be very endangered.

Limnanthes floccosa ssp. pumila, an annual, is glabrous throughout, about four inches tall, with petals and sepals the same size. The petals are white, one-fourth to one-third inches long, greenish at the base on the inside and lined with green longitudinally. The small leaves are pinnate with linear lateral divisions. It blooms in March and April.

184
Slender meadow-foam
(*Limnanthes gracilis* Howell *var. gracilis*).
Meadow-foam Family (*Limnanthaceae*).
Found in Douglas, Jackson and Josephine counties in Oregon, in areas very wet in early spring, most often in serpentine soil, this meadowfoam is rare but currently stable.

Limnanthus gracilis var. gracilis is a glabrous annual, three to four inches tall, with very small pinnate leaves at the base of branching stems, and along the stem near the base. The stems may be reddish, the sepals bright green and about one-half the length of the petals. The petals, about one-half inch long, are white to slightly pink near the tips, yellowish near the base on the inside and veined with green from base to tip. It blooms from March through May.

185
Alp lily
(*Lloydia serotina* [L.] Salisb. ex Reichenb.).
Lily Family (*Liliaceae*).
This small, high elevation lily is found in Oregon only near the summits of Saddle Mountain in the north Coast Range, in the Wallowa Mountains in northeast Oregon and on Steens Mountain in the southeast part of the state. Although it is found in the Olympic Mountains of Washington, the Rocky Mountains from Alberta to New Mexico, in Alaska, Europe, and Asia, it is rarely seen in Oregon. It grows in rocky crevices, cliffs, and summits and on gravelly ridges and slopes.

Lloydia serotina is about eight inches high. The leaves are very narrow, generally not much taller than the flowering stem; there are both basal and cauline leaves. The six tepals, each one-half inch long, are white with greenish veins on the inside; the outside is tinged with purple toward the base. It blooms from May to June.

186 187

186
Bradshaw's lomatium

(*Lomatium bradshawii* [Rose] Math. & Const.). Parsley Family (*Apiaceae* or *Umbelliferae*).
This species, endemic to and once widespread in the wet, open areas of the Willamette Valley of western Oregon is limited now to a few sites in Lane, Marion, and Benton counties. Most of its habitat has been destroyed by land development for agriculture, industry, and housing. It is now endangered throughout its range.

Lomatium bradshawii grows from eight to twenty inches tall, is glabrous, and has chiefly basal leaves that are divided into very fine, almost thread-like, linear segments. The yellow flowers are subtended by green bracts divided into three's or doubly divided into three's. This bract arrangement differentiates it from other lomatiums. It is in bloom in April and May. Its fruits appear in late May and June. They are oblong, about one-half inch long, thick-winged along the margin, and have thread-like ribs on the dorsal surface.

Ternately divided bracts beneath flowers

187
Columbia lomatium or Purple leptotaenia

(*Lomatium columbianum* Math. & Const.). Parsley Family (*Apiaceae*).
This big showy purple lomatium is a regional endemic known only at the east end of the Columbia River Gorge in Hood River and Wasco counties in Oregon and north to Yakima County in Washington. It grows on dry, open rocky slopes.

Lomatium columbianum is large and almost shrublike, glabrous and glaucous. It grows from twenty to forty inches tall. The leaves are stout and many times divided into narrowly linear segments. The flower stems terminate in an umbel of ten to twenty flower heads each subtended by several long, narrowly triangular, pointed, reddish bracts. It is one of the few purple-flowering lomatiums. It blooms in March and April and produces a fruit about an inch long, oblong in shape, with thick wings and inconspicuous dorsal ribs.

188 189

Lanceolate undivided bracts beneath flowers

188
Agate Desert lomatium
(*Lomatium cookii* Kagan).
Parsley Family (*Apiaceae*).
This recently discovered species, presently known only from the "Agate Desert" of Jackson County in southern Oregon is considered to be rare and endangered throughout its range. As information is scant concerning this new species, study and research are continuing.

Lomatium cookii grows to about twelve inches tall, is light green in color, glabrous and has stems which are grooved quite deeply longitudinally. The erect basal leaves are about as tall as the flowering stems. They are several times divided into very narrow, linear segments. There are no involucral bracts (whorls of modified leaves at the base of the umbel), but there are involucel bracts (bractlets subtending each head of flowers within the umbel) by which this species is differentiated from others. About fourteen in number, these bractlets are long, narrow, broadest in the the middle and pointed at the tip. The flowers are yellow, five-petalled, and clustered tightly, with stamens well-exserted. It blooms in March and April.

189
Greenman's lomatium
(*Lomatium greenmanii* Mathias).
Parsley Family (*Apiaceae*).
A very rare species known only from one site, on the summit of Mt. Howard in the Wallowa Mountains of northeastern Oregon. Collected originally by Cusick in 1890, it was not rediscovered until 1975. It is considered rare and endangered throughout its range. It grows in an alpine or subalpine situation on windswept ridges and knolls, in moist meadows or lightly forested areas.

Lomatium greenmanii is a glabrous, dwarf perennial, two to seven inches tall; the leaves are mostly basal, pinnate to bipinnate, having leaflets nearly one-half inch long and less than one-eighth inch wide. The flowers are yellow, arranged in small, compact heads supported by a few narrow bractlets. The oval-shaped fruits are glabrous, about one-eighth inch long, two-thirds as wide as long, with narrow dorsal wings and slightly wider marginal wings. It flowers from late June through August.

190 191

190
Howell's lomatium
(*Lomatium howellii* [Wats.] Jepson).
Parsley Family (*Apiaceae*).
This rare species is the only one in the subgenus, *Euryptera*, of the genus *Lomatium*. It is endemic to the dry serpentine slopes in the Illinois and Chetco drainages of Josephine and Curry counties in Oregon and adjacent northern California.

Lomatium howellii is glabrous, with grayish-green foliage, grows eight to fifteen inches tall, its stems are usually without leaves. Its basal leaves are arranged in threes, then pinnate. The leaflets are broad, almost round, up to an inch in length, very thick and tough and sharply toothed or sometimes lobed. There are from six to sixteen rays in an umbel, each flowering head subtended by a few linear bractlets. The small flowers are yellow, fading to white as they age. The fruit is nearly orbicular, notched at both ends, about one-fourth inch long, with peripheral "wings" half the width of the body. It blooms in May and June.

191
Smooth desert parsley
(*Lomatium laevigatum* [Nutt.] Coult. & Rose).
Parsley Family (*Apiaceae*).
This rare desert parsley is known only from the eastern end of the Columbia River Gorge in Wasco County in Oregon, and Klickitat County in Washington. It grows on basaltic cliffs.

Lomatium laevigatum is glabrous, ten to twenty inches high, has leaves that are divided into three then pinnate with leaflets parted into uniform segments. There are no bracts or bractlets subtending the umbels or the flower heads within the umbel. The flowers are yellow, the fruits elliptical, almost one-half inch long and broad-winged. It blooms from March to May.

192
Colonial luina or Creeping silverback
(*Luina serpentina* Cronq.).
Composite Family (*Asteraceae*).
This striking plant is known only from a few sites on the steep, rocky hillsides above Fields Creek near Dayville in Grant County, Oregon. Although originally thought to be endemic to serpentine, it actually grows on non-serpentine soil in the vicinity of serpentine rock. It is rare throughout its narrow range.

Luina serpentina is a perennial, six to ten inches tall, which forms large silvery mats more than six feet in diameter. The sessile leaves are narrow, entire, lance-shaped, and densely white-woolly underneath, lightly so above. The flowers are bright yellow disk-flowers with long exserted yellow stamens. Each flower head is surrounded by an involucre of about ten to seventeen narrow, straight, parallel, very white-woolly bracts, about one-third of an inch long. It blooms in July.

Close-up of flowers

193
Mt. Ashland lupine
(*Lupinus aridus* Dougl. ex Lindl. *ssp. ashlandensis* Cox).
Pea Family (*Leguminosae* or *Fabaceae*).
This endangered subspecies is found only on the dry granitic outcrop of rock at over 7000 feet elevation on the summit of Mt. Ashland.

Lupinus aridus is a perennial, growing in clumps to a height of five inches, its leaves ascending, with five to seven leaflets, each about three-quarters of an inch long. The flowering stems, three inches long, terminate in a dense raceme, one and one-half to three inches tall, of many purple flowers. It is densely woolly throughout except for the flower petals. It blooms in July and August.

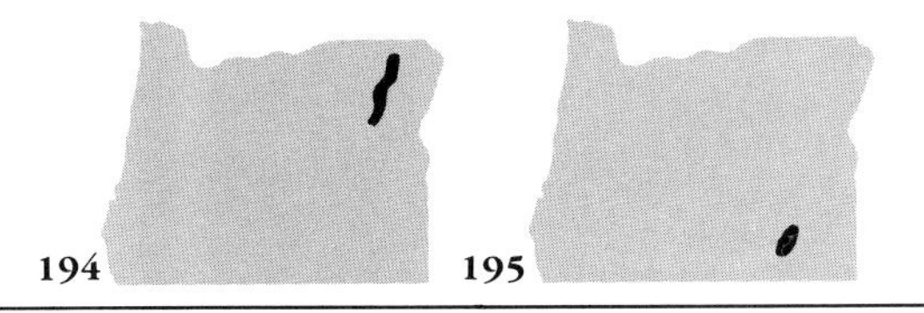

194
Blue Mountain lupine
(*Lupinus burkei* Wats. *ssp. caeruleomontanus* Dunn & Cox).
Pea Family (*Fabaceae*).
This rare lupine, found only in the Blue Mountains in eastern Oregon, was first described in 1973. Although limited in both range and numbers it is considered to be currently stable but in need of continued monitoring.

Lupinus burkei ssp. caeruleomontanus grows to thirty inches in height, and has purple to purplish-white flowers. It is distinguished by its large, long-stemmed leaves which have seven to eight leaflets as long as three inches and as wide as one and one-half inches, with their greatest width being closer to the rounded or obtusely pointed tip. It blooms in June and July.

195
Small-leaved lupine
(*Lupinus lyallii* Gray *ssp. minutifolius* [Eastw.] Cox).
Pea Family (*Fabaceae*).
Also known as *Lupinus lobbii* Gray *ssp. minutifolius.* This subspecies is known only from a small area along the summit ridges of Steens Mountain in Harney County, where it is found growing in rocky or gravelly areas at elevations from 7500-9700 feet. It is currently stable throughout its range, but its range is very small.

Lupinus lyallii ssp. minutifolius is covered with an appressed pubescence giving its foliage a grayish-green appearance. Each leaf consists of from five to seven elliptical leaflets arranged like the spokes of a wheel at the tip of the stem. The leaflets are quite small, one-third of an inch long and one-eighth of an inch wide. The flowers, in a rather tight raceme about one and one-half inches long, are purplish-blue and white. It blooms from July into September.

196
Sabine's lupine
(*Lupinus sabinii* Dougl. ex Hook.).
Pea Family (*Fabaceae*).
This rare lupine is found in the Blue Mountains of Umatilla, Union, and Morrow counties, and has been reported as far south as Steens Mountain in Harney County. Its range extends north into Walla Walla County, Washington. It grows in somewhat dry areas of coniferous forests and prairies, and has recently been found on open sagebrush land. It is listed as limited in abundance but stable in Oregon; threatened in Washington.

Lupinus sabinii is a perennial growing to 48 inches tall. It has leafy stems and its foliage is covered with rather short, stiff, appressed hairs. Each leaf has nine to eleven lance-shaped leaflets that may be up to six inches long and one inch wide. The flower racemes are six to twelve inches high with numerous, but not crowded, bright yellow pea-shaped blossoms. The pods, nearly two inches in length, appear silky with long, appressed shiny hairs. It blooms May to July.

197
Kincaid's sulfur lupine
(*Lupinus sulphureus* Dougl. *var. kincaidii* [Smith] Hitchc.).
Pea Family (*Fabaceae*).
This is one of three varieties of *Lupinus sulphureus* found in Oregon. It is known in the Willamette Valley and south into Douglas County, Oregon, with a disjunct population reported in Lewis County, Washington. It is considered to be a threatened species in Washington, but is still on the Review List in Oregon. It is undetermined here as to whether sites out of the locally known range are of this variety.

Lupinus sulphureus var. kincaidii is a beautiful lupine with its unusual appearing leaves, eye-catching from a distance. The leaves are oval-palmate with very narrow leaflets that are deeply creased at the mid-line, glabrous on the upper surface, but covered with silvery-white hairs underneath. The reddish stems and the sepals are also covered with silvery hairs. The small, purplish-blue, pea-shaped flowers are in loose racemes four to eight inches tall. (A yellow variation of this variety has been reported at Fern Ridge). The flowers have very short, densely hairy pedicels. The blooming period is from May to July.

198
Inch-high lupine
(*Lupinus uncialis* Wats.).
Pea Family (*Fabaceae*).
A species of the Nevada desert, this very tiny inch high lupine is found only in small isolated areas of southern Harney and Malheur counties in Oregon, and recently, in Owyhee County, Idaho. It grows in sandy, pebbly soil.

Lupinus uncialis forms a round mat one inch to two and a half inches across, about one-half to one inch tall. Its stems divide at ground level into two or three whitish, thickened, prostrate branches, these again branching profusely. The minute leaves are palmate and are divided into five very hairy leaflets that are usually less than one-quarter inch long. The flowers are few and solitary in the leaf axils. The corolla, about one-fourth inch long, is creamy-white with purple margins and tips on the petals. It blooms in May and June.

Close-up of minute flower and leaves

199
Alaskan club-moss
(*Lycopodium sitchense* Rupr.; also known as *Lycopodium sabinaefolium* Willd. *var. sitchense* [Rupr.] Fern.).
Clubmoss Family (*Lycopodiaceae*).
This is a spore-producing plant belonging to the Phylum *Pteridophyta* (ferns and fern allies). It is found in moist alpine country, usually at timberline or above, from Oregon to Alaska, across America to New England and in eastern Asia. Its occurrence here is restricted to this habitat and in Oregon it is known only from the high Cascades and Wallowa Mountains.

Lycopodium sitchense has stems which creep along the ground. From these stems arise erect, dichotomous, aerial branches. The leaves of the branches are five-ranked, linear, close to the stem, sharply pointed, one-eighth inch long, and thickened. The spore-producing cones at the ends of the aerial branches are round in cross-section, and almost an inch in length. It was once reviewed for rarity, and though it is seldom seen and little known, it was found to be stable in Oregon. It produces strobili (cones) in the fall.

Habitat scene

200
Fringed water-plantain
(*Machaerocarpus californicus* [Torr.] Small; also known as *Damasonium californicum* Torr.).
Water-plantain Family (*Alismaceae*).
This is a perennial water plant which grows in shallow ponds and sloughs. Although its range extends from the Washington side of the Columbia Gorge, south along the east side of the Cascades into California, Nevada, and southwestern Idaho, it is known from very few sites in Oregon. It was placed on the Review List in Oregon in 1983 for further study.

Machaerocarpus californicus has lanceolate to ovate leaves, with entire margins, two to three inches long, that are on stems longer than the blade of the leaf and equal in length to the stems of the inflorescence. They are usually lax, floating on the surface of the water. The leafless flowering stems are topped with one or two whorls of flowers which have three broad, white to pinkish petals, less than one-half inch long, that are deeply fringed along their outer margin, and three broadly ovate, greenish sepals. There is often a touch of yellow at the base of each petal. The flowers are perfect, with six stamens and usually six pistils. The achenes, small one-seeded fruits, are divergent, and sharply contracted to a long beak. It blooms from April into August.

201
Torrey's malacothrix
(*Malacothrix torreyi* Gray).
Composite Family (*Asteraceae*).
A species of southeastern Oregon, Harney and Malheur counties, south into Nevada and east to Idaho and Utah. It is found in dry sandy or gravelly places. In Oregon this plant is considered to be stable at present, but needs periodic monitoring.

Malacothrix torreyi is a somewhat decumbent annual, with stems up to twelve inches in length. Its herbage is generally glabrous, but may have some glandular-tipped hairs. The leaves, both basal and along the stem, are from one and a half to two and a half inches long and are deeply lobed along the sides. The lobes are sharp-pointed and unequal in length. The lower leaves have winged stems while those above are clasping. The flowers are yellow, several to a stem, and ligular. The outer ligules are one-quarter to one-third of an inch in length. It blooms from May to July.

202
White or Oregon meconella
(*Meconella oregana* Nutt.).
Poppy Family (*Papaveraceae*).
This rarely seen very small annual poppy is found sporadically at low elevations from Vancouver Island to California. In Oregon it is known from sites in the Willamette Valley and the Columbia Gorge. It grows in open areas that are wet in the spring, often hidden in the grass with which it is growing. It is considered to be threatened in Oregon, and sensitive in Washington.

Meconella oregana grows from one inch to five inches tall with very slender simple or branched stems. The plant is glabrous throughout. The basal leaves are spatulate or obovate, rounded but widest toward the tip. The stem leaves are smaller and clasp the stem. The flowers are single at the top of each stem, have three sepals, four to six petals (usually six), that are white, with a yellowish-green base, and are less than one-quarter inch long. The stamens are much shorter than the pistil. The seed pods are erect, linear, sometimes twisted, about three-quarters of an inch long, and attached to the narrow spreading rim. Its genus name is Greek meaning a "little poppy". It blooms in April and May.

Close-up of minute basal leaves

203
Packard's mentzelia
(*Mentzelia packardiae* Glad).
Blazing Star Family (*Loasaceae*).
Discovered in the mid 1970's in eastern Malheur County, it is an extremely restricted endemic growing only in a very specific layer of volcanic ash. It is considered to be rare and endangered throughout its range.

Mentzelia packardiae is an annual growing erectly to a height of sixteen inches, has linear, entire basal leaves, and sessile, clasping stem leaves. The foliage is covered with fine whitish hairs giving it a gray-green appearance. There are five triangular, pointed sepals, and five broadly elliptic petals up to five-eighths of an inch long, which are a shiny yellow with a touch of bronze at their base. The yellow stamens are numerous, over sixty, and about three-eighths of an inch long. The seeds are irregularly angled. It blooms in May and June.

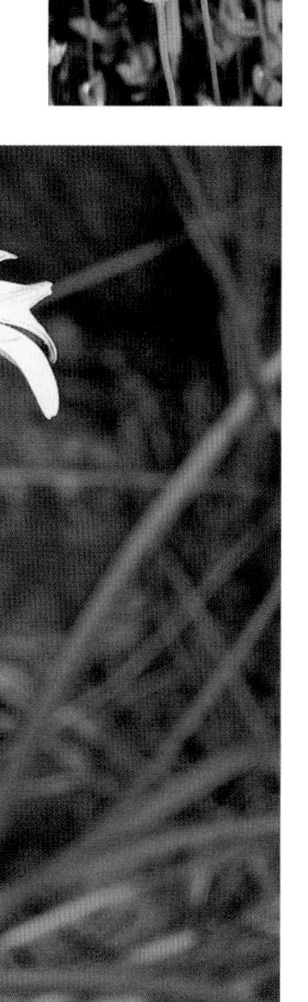

Flowers closed on a shady day

204
Timwort

(*Microcala quadrangularis* [Lam.] Griseb.).
Gentian Family (*Gentianaceae*).
This tiny plant exists over a wide range, but has been found in only a few widely separated locations in the upper Willamette and the Umpqua valleys in Oregon, in northeast California, and in Peru in South America. It is very rare and threatened in Oregon.

Microcala quadrangularis grows from one to four inches tall; its glabrous stem is thread-like, often branching into two or three near the base. It has a few basal leaves, each about one-third of an inch long, and usually two pairs of smaller, opposite stem leaves. The single flowers at the top of each stem are bright yellow, have four petals that are about one-third of an inch long. They apparently do not open except on sunny days. The "quadrangular" calyx tube, one-quarter of an inch in length, has four prominent ridges with sharp-pointed tips. It blooms from May to June.

Basal leaf

205
Howell's microseris

(*Microseris howellii* Gray).
Composite Family (*Asteraceae*).
This plant is a rare regional endemic found on moist serpentine slopes in the Siskiyou Mountains of southwest Oregon. It appears to be restricted to the drainage of the Illinois River, though may possibly exist in adjacent northern California. It is threatened throughout its range.

Microseris howellii is a perennial herb with a slender stem rising singly from a taproot to a height of about twenty inches. The leaves, linear or pinnatified with long, slender lobes are chiefly basal, up to twelve inches long. The foliage is generally glabrous, or lightly covered with a rough scale-like pubescence. The flower heads are narrow, nodding before blooming, have fifteen to twenty-five yellow ray flowers. It blooms from April to July.

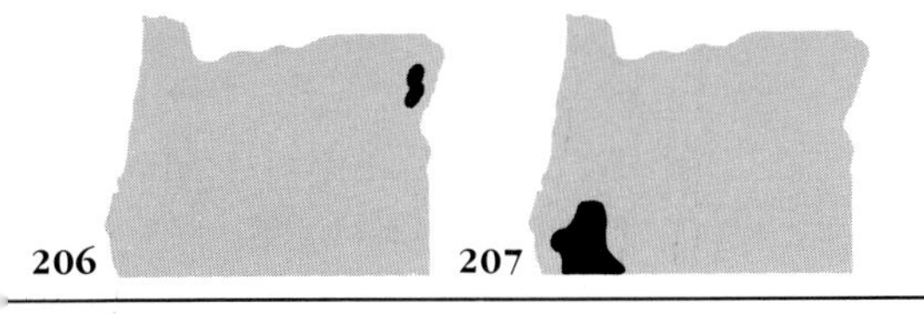

206
Hill monkeyflower or Bank monkeyflower
(*Mimulus clivicola* Greenm.).
Figwort Family (*Scrophulariaceae*).
This rare monkeyflower grows on moist, often quite steep, slopes at elevations of five to six thousand feet elevation, from the Snake River Canyon, Baker and Wallowa counties, Oregon, to Washington and northern Idaho. It is threatened if not endangered in Oregon.

Mimulus clivicola is an attractive, small (two to eight inches tall) plant with big purplish-pink flowers. The smaller plants tend to have but one flower which seems much too big for the plant. Larger plants may have six to eight flowers. The leaves are elliptic, entire or sometimes sharply toothed, up to an inch in length. The calyx is stongly angled, two-fifths of an inch long, with sharp teeth. The plants are covered with a glandular pubescence. The flowers, three-quarters of an inch long, have purplish-pink lobes, shading to yellowish-pink with some red spots in the throat, to a distinct yellow in the narrow, elongated tube. The flowers are bilabiate, and tend to persist after blooming, which is uncommon for *Mimulus*. It blooms in June and July.

Lateral view showing yellow flower tubes, angled calyces and persistant corollas

207
Douglas' monkeyflower
(*Mimulus douglasii* [Benth. in DC.] Gray).
Figwort Family (*Scrophulariaceae*).
Also called, very properly, "purple mouse ears". It grows in moist soil or gravelly places, usually on serpentine, in Douglas, Curry, Josephine, and Jackson counties of southwest Oregon, south to central California. It is rare and threatened in Oregon but more common in California.

Mimulus douglasii grows to two and one-half inches tall; the foliage is somewhat villous with glandular hairs. The leaves, up to an inch long, are numerous and crowded at the base. The one to four flowers have deep purple tubes, one to two inches long, which abruptly flare into a broad throat streaked with darker purple. The upper petal lobes are large, rounded and rose-purple; the lower lobes are almost non-existent. The four stamens, in two unequal pairs, are slightly exserted and have bright yellow anthers. The calyx, almost one-half inch long, is tubular, greatly dilated, with prominent, dark green verical ridges. It blooms from March into May.

Habitat view

208
Hepatic monkeyflower
(*Mimulus jungermannioides* Suksd.).
Figwort Family (*Scrophulariaceae*).
This rare species originally found near Bingen on the Washington side of the Columbia Gorge, is now believed to be extinct in Washington, but can still be found at a few scattered sites in Wasco, Sherman, Gilliam, and Umatilla counties, Oregon. It grows chiefly in the crevices of damp rock walls.

Mimulus jungermannioides is a perennial with long thread-like stems which send off roots at the nodes into the crevices of the rock on which it grows. The stems may be four to twelve inches long, and are very glandular-villous, as are the leaves. The leaves are oval in shape, toothed along the margins, rounded to the stem which may be as long as the leaf, about three-quarters of an inch. The flowers are about an inch long, are yellow with a few red dots on the lower petals. The throat is wide open, very hairy within; the lobes are rounded, spreading widely. The tube is funnelform, narrow, and about twice the length of the calyx. The anthers are glabrous. Starting early in July it blooms most of the summer.

209
Kellogg's monkeyflower
(*Mimulus kelloggii* [Curran ex Greene] Gray).
Figwort Family (*Scrophulariaceae*).
This California species was first identified in Oregon in 1977 and is still found in just a few places in Douglas and Jackson counties in southwest Oregon. It is rare and threatened in Oregon. It grows in open places of coniferous forests in clay-like soil.

Mimulus kelloggii is an annual which reaches only four or five inches in height. It is covered with gland-tipped hairs, causing it to be sticky, and collect dirt and debris which may blow past. The calyx is about one-half of an inch long, with glandular-pubescent ridges, the uppermost being the longest. The flowers, few on a stem, have a long white tube which flares into a deep purple throat with five lips or lobes (two upper and three lower). The lips become paler in color toward the periphery and have two yellow ridges on the lower lip at the entrance to the throat. It blooms from April to June.

210 211

210
Stalk-leaved monkeyflower
(*Mimulus patulus* Penn.).
Figwort Family (*Scrophulariaceae*).
An annual monkeyflower, very glandular and hairy, growing generally on damp rock walls in Wallowa County, Oregon and Whitman County in southeast Washington. It is rare and threatened in Oregon.

Mimulus patulus has slender, erect stems, about eight inches tall. The leaves, about one-half inch long, are rounded to cordate in shape, and have margins that are somewhat undulate. They are palmately three-veined, and are abruptly contracted to a slender stem, which is one to two times the length of the leaf blade. The calyx is angled and about one quarter inch long with very short lobes. The corolla is yellow, has rounded lobes, and a narrow, bell-shaped, open throat, with two low, slightly pubescent ridges that are sometimes dotted with red. The anthers are glabrous. It blooms in June and July.

211
Pulsifer's monkeyflower
(*Mimulus pulsiferae* Gray).
Figwort Family (*Scrophulariaceae*).
This seldom seen annual monkeyflower is found in scattered populations east of the Cascades from Washington to California. It grows in moist open areas, and is often found in association with *Mimulus pygmaeus* in the southern portion of its range. It is presently stable in Oregon but needs to be monitored occasionally.

Mimulus pulsiferae grows only to about six inches tall. Its branches are opposite and often curved in the form of a candelabra. Its herbage is very glandular and puberulent. The yellow, tubular corolla is less than one-half of an inch in length, with its throat gradually widening to the short, spreading lobes. There is sometimes a red spot in the throat. The calyx is narrow and tubular, ridge-angled with short, triangular-acute lobes. The leaves, slightly more than one-half inch long, are elliptic to oblong on short stems, and may or may not be toothed. It blooms from April to June.

212 213

Pygmy monkey-flower shown with pulsifer's monkey-flower (left)

212
Pygmy monkeyflower
(*Mimulus pygmaeus* Grant).
Figwort Family (*Scrophulariaceae*).
This rare species found only in Jackson County, Oregon, and in Lassen, Modoc, and Plumas counties in California is considered to be rare and endangered throughout its range. It is so limited in its populations and present in such small numbers that it is seldom reported. It prefers moist flats or hillsides at elevations of 4000-5000 feet, and is often found associated with *Mimulus pulsiferae.*

Mimulus pygmaeus is an annual, and as its name implies, very small, growing only about one-half inch tall, the whole plant being less than an inch across. The leaf blades are linear to oblanceolate, about one-third of an inch long, and glandular pubescent. The calyx, less than one-quarter inch long, has unequal lobes, and is also glandular pubescent. The corolla, yellow to slightly purplish, is about one-quarter to one-third inch long. Its throat, slightly shorter than the tube, is bell-shaped; the orifice is open but quite flattened as the flaring, rounded lobes are wider horizontally than vertically. There may be red dots on the inner surface of the lower petal. It blooms May and June.

213
Tri-colored monkeyflower
(*Mimulus tricolor* Hartw. ex Lindl.).
Figwort Family (*Scrophulariaceae*).
This flower was once common in the Willamette Valley, covering the fields purple in spring. Land development has critically reduced the populations to just a few plants in scattered locations in Marion, Benton, Linn, Klamath, and Lake counties in Oregon and south into central California. It is now rare and threatened in Oregon. It grows at low elevations in clay soil, especially preferring vernal pools which dry up later in the season.

Mimulus tricolor is hairy and sticky-glandular, and is very small, perhaps reaching five inches in height. Its leaves are about one inch long and its corolla one to one and one-half inches long. The yellowish tube of the corolla bends slightly into the throat which is dark purple outside, white with two yellow patches and several dark purple spots within. The throat spreads into five lobes which are light purple with dark purple median spots. It blooms from late May into June.

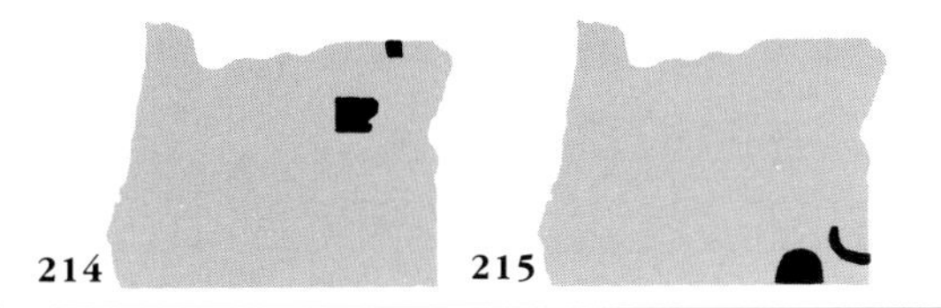

214
Washington monkeyflower
(*Mimulus washingtonensis* Gand.).
Figwort Family (*Scrophulariaceae*).
This plant grows in moist areas, chiefly in streambeds, in dry areas of central Oregon, east of the Cascades, and into southern Washington. The species as a whole is quite common throughout its range, however *var. washingtonensis*, found only in Grant and Wallowa counties in eastern Oregon, is very rare and threatened throughout its range.

Mimulus washingtonensis is diminutive, rarely over six inches tall, very hairy and glandular. The ovate, denticulate leaf blades are about one-half inch long, oppositely paired, and purple-tinged. The calyx, one-third inch long, is prominently angled with pointed lobe-tips. The flowers are bright yellow, about three-quarters of an inch long. The throat has brownish-red spots and is pubescent on the two ridges within. It blooms from May into the summer, or until its habitat dries up.

215

Desert four-o'clock or Wishbone bush
(*Mirabilis bigelovii* Gray *var. retrorsa* [Heller] Munz).
Four-o'clock Family (*Nyctaginaceae*).
In Oregon this species is found only southeast of Steens Mountain in Harney County and in the Owyhee Canyon in Malheur County. It is more common to the south in Nevada, Arizona, and southern California. It is currently stable in Oregon but bears watching. It grows on rocky slopes below basaltic cliffs.

Mirabilis bigelovii var. retrorsa is a much branched herb with a shrubby base, up to 26 inches tall. The upper part of the plant is covered with harsh, dense, retrorse, glandular hairs. The base of the plant is nearly glabrous. The leaves are broad, lance-shaped to nearly orbicular, a little over an inch in length, with short stems. The flowers, several in a cluster at the tip of the stems, are subtended by a whorl of modified leaves or bracts called an involucre. They are white or pinkish and consist of sepals, rather than petals, that are less than one-half inch in length and bilobate at the tips. It blooms in June.

Close-up of flower

216
Macfarlane's four-o'clock
(*Mirabilis macfarlanei* Const. & Roll.).
Four-o'clock Family (*Nyctaginaceae*).
One of Oregon's rarest wildflowers, this species known only from the vicinity of the Snake and Imnaha river canyons has been formally listed by the Federal Government as endangered in both Oregon and Idaho. It is believed to be on the brink of extinction. This *Mirabilis* grows on steep, sunny slopes of talus covered with a thin layer of soil.

Mirabilis macfarlanei is a perennial arising from a deep-seated root which forms a clump with many branches. The shiny, succulent, nearly-sessile leaves are opposite. Those at the base broadly ovate and rounded at the tip, two to three inches long; the upper ones are pointed and smaller. A purplish-colored whorl of bracts at the tip of the stem, or in leaf axils, encloses several flower buds which seem to bloom only one or two at a time. The flower is a rich magenta, very showy funnel-like blossom of sepals, three-quarters to one inch in length; there are no true petals. The fruits, about one-third of an inch long, are elliptical in shape, circular in cross-section, have a wrinkled surface, and are obscurely ten ribbed. It blooms in May and early June.

217
Siskiyou monardella
(*Monardella purpurea* How.).
Mint Family (*Lamiaceae*).
Found only on serpentine soil on high open rocky slopes in Curry and Josephine counties in Oregon and in Del Norte, Humbolt, and Siskiyou counties in California, it is a regional endemic considered to be rare and threatened in Oregon.

Monardella purpurea forms clumps of purplish puberulent branches up to six inches high, with narrow ovate leaves about one inch long which are shiny above and minutely puberulent beneath. The flowers, numerous in tight clusters within the bracts, are a rich rose-purple, not quite an inch in length well-exserted from the calyx. The bracts of the flower heads are three-quarters of an inch long, ovate, rounded at the tip, puberulent, purple-tinged. The flower heads are about an inch across. The calyx, too, is hairy and purplish-colored, less than one-half inch long, with narrow, pointed sepals. It blooms from June to September. It is sympatric with the common, very similar species, *Monardella odoratissima* Benth., but does not have its smell of mint.

218
Lobb's nama
(*Nama lobbii* Gray).
Waterleaf Family (*Hydrophyllaceae*).
A species of California, known in Oregon only in Jackson and Klamath counties, it grows in open pine forests in sandy or rocky soil. It was originally discovered in Oregon on the slopes of Mt. McLoughlin, then known as Mt. Pitt, in 1899. Not seen for many years, it was thought to be extinct in Oregon until rediscovered recently in several sites east of the mountain. It is presently rare but stable in Oregon; more common in California.

Nama lobbii is a low, spreading shrubby-based plant with stems two to twelve inches tall. Its leaves are about an inch in length, broadly linear in shape, and covered with dense, white woolly hairs. The flowers are stemless, rising from leaf axils near the end of the stems. The corolla is purple, about one-third of an inch long, and funnel-shaped, with five flaring lobes. The calyx, half the length of the corolla, has narrowly linear segments. The globose capsule contains about twelve small, dark brown seeds. It blooms from July to September.

Close-up of flowers

219
Wolf's evening primrose
(*Oenothera wolfii* [Munz] Raven, Dietrich and Stuble).
Evening Primrose Family (*Onagraceae*).
This rare plant, found chiefly in sandy soil on bluffs above the ocean beach, is known only from a few sites in Curry County, Oregon, and in Del Norte and Humboldt counties, California. It is a candidate for federal listing as being endangered.

Oenothera wolfii is a perennial, or sometimes biennial plant, growing erectly from twenty to sixty inches tall. The stems are coarse, sometimes reddish or reddish-green in color, and leafy nearly to the top with sessile, lanceolate leaves from one to five inches long. The basal leaves, which form a rosette, are four to seven inches long, and pointed. The entire plant is covered with coarse, stiff hairs. The flowers are densely arranged among the smaller leaves near the top of the stem. The four petals are about an inch long, yellow, turning to orange in age. The sepals are long, triangular, ciliate, and reflexed backward against the stem. The stigma is divided into four spreading, linear lobes. The inflorescence is glandular and sticky, and the fruits are angled and tapering. It reportedly blooms from June to October.

Close-up of flower

220
Broad-scaled orthocarpus
(*Orthocarpus cuspidatus* Greene).
Figwort Family (*Scrophulariaceae*).
This species is found on open slopes of the high summits in the Siskiyou Mountains in southern Jackson County, Oregon and adjacent Siskiyou County in northern California. It is currently on the Watch List in both Oregon and California as being stable but in need on continued monitoring.

Orthocarpus cuspidatus has a stout, erect stem, sometimes branched near the top, which may grow to about fourteen inches tall. The leaves, both basal and along the stem, are one to two inches long, and are divided into narrow, linear segments. The bracts are abruptly different. They are broad, ovate, less than three-quarters of an inch long, with a pair of spreading, narrow lobes on either side, and overlap closely like shingles. They are green, tinged with purple toward the tip. The flowers, much longer than the bracts, consist of an upper purple lip called the galea that is long, narrow, arching, pubescent to the tip, and a lower lip which is white, inflated, shallowly divided into three lobes, and about three-sixteenths of an inch shorter than the galea. It blooms from June to August.

221
Mt. Hood grass-of-parnassus
(*Parnassia fimbriata* Konig. *var. hoodiana* Hitchc.).
Saxifrage Family (*Saxifragaceae*).
This variety, once thought to grow only in the vicinity of Mt. Hood, is now known from Clackamas County, Oregon to Skamania County, Washington. It is listed as sensitive (vulnerable and declining) in Washington, but appears to be more stable in Oregon. It grows at fairly high elevations in marshes, bogs, wet mountain meadows and along streambanks.

Parnassia fimbriata var. hoodiana grows ten to sixteen inches tall, has glabrous, ovate to kidney-shaped entire basal leaves that are from one inch to one and one-half inches in diameter. The flowers are solitary on leafless stems, have five rounded, white petals that are marked with greenish veins. The petals have fringed (fimbriate) lateral margins. The five stamens which are flared from the base of the superior ovary, have white filaments and yellowish anthers. Protruding also from the base of the ovary, between the filaments, are five yellowish sterile stamens called staminodia. In *var. hoodiana* these are broad and fused at the base, dividing about half-way into six to eight delicate, filament-like appendages, with small, globe-like heads. It blooms from July to September.

222
Bracted lousewort
(*Pedicularis bracteosa* Benth ex Hook. *var. pachyrhiza* [Penn.] Cronq; also known as *Pedicularis pachyrhiza* Pennel, rather than as the variety).
Figwort Family (*Scrophulariaceae*).
This special plant is found only in the Wallowa and Blue Mountains of northeast Oregon and southeast Washington. It grows in coniferous forests and damp thickets.

Pedicularis bracteosa var. pachyrhiza is a perennial which grows to a height of over three feet, from a thick, tuberous root. The lower part of the plant is glabrous; the flowering head quite hairy. The leaves, up to six inches long, are pinnatifid with long, narrow, toothed segments. The calyx, with the uppermost lobe being the smallest, is not glandular. The corolla is greenish-yellow, about three-quarters of an inch long, has an upper hood-shaped lip called the galea. The lower lip is only slightly shorter than the upper. The bracts in the inflorescence are about as long as the flowers. It blooms from June to August.

223
Indian warrior
(*Pedicularis densiflora* Benth ex Hook.).
Figwort Family (*Scrophulariaceae*).
A striking plant because of the bright red color of its flowering head, this California species is found in Oregon only in the Siskiyou Mountains. It grows in gravelly or sandy soils in oak or pine forests.

Pedicularis densiflora has a stout stem up to ten inches tall; the leaves are pinnatifid with ten to twenty pairs of leaf segments that are deeply cleft or toothed into more segments. The flowers are one to one and one-half inches long. The upper lip is cylindric, arched, and without hairs; the lower lip is much shorter with broad, fringed, somewhat yellowish lobes. It blooms from March to June.

Close-up of flowers

224
Hedgehog cactus
(*Pediocactus simpsonii* [Engelm.] Britt & Rose *var. robustior* [Coult.] L. Benson).
Cactus Family (*Cactaceae*).
The only barrel cactus native to Oregon, it is found in the dry areas east of the Cascades, from eastern Washington south into Nevada, and east into Utah, Wyoming and Colorado. This very attractive plant, much sought after by collectors, is rare but currently stable in Oregon, however, its populations need to be monitored.

Pediocactus simpsonii var. robustior grows singly or in clusters, and is three to five inches in diameter and about as tall above the ground. It has numerous whorls of sharp, stout white and brownish-purple spines that are up to an inch long. The flowers, in a crowded ring at the top center of the spiny globe, are usually a rosy-pink color, sometimes white, each one up to an inch in diameter. They may be found in bloom from April to early May.

225
Barrett's penstemon
(*Penstemon barrettiae* Gray).
Figwort Family (*Scrophulariaceae*).
This striking beardtongue, named for the wife of a Dr. Barrett in Hood River who discovered it in the 1880's, is endemic to a limited area of the Columbia River Gorge, Hood River and Wasco counties in Oregon, and Klickitat County in Washington. Attractive to collectors, this plant is threatened throughout its range. It is most commonly found growing on basaltic cliffs at low elevations.

Penstemon barrettiae has grayish-green, evergreen leaves up to an inch and a half long which are toothed along the margins. The stems grow to sixteen inches long and are topped with a raceme of large, rose-purple flowers up to an inch and a half in length. The anthers are densely woolly. It blooms from April through June.

226 227

226
Davidson's penstemon

(*Penstemon davidsonii* Greene *var. praeteritus* Cronq.).
Figwort Family (*Scrophulariaceae*).
This penstemon is found at high elevations on Steens Mountain and in the Pueblo Mountains of southeast Oregon and in northern Nevada, where it grows on cliffs, rocky outcrops and talus slopes in basaltic or granitic soil. It is limited in abundance throughout its range but is currently stable.

Penstemon davidsonii var. praeteritus is a showy plant with bluish-lavender bilabiate flowers, one and one-half inches to nearly two inches long. The upper lip has two lobes; the lower has three. The anthers have a tuft of hair referred to as a "coma", and are covered with pores that discharge pollen. The leaves are thickened, elliptical in shape, and glabrous. A creeping shrub, it tends to form mats over the rocks. It blooms from June to August.

227
Crested tongue penstemon

(*Penstemon eriantherus* Pursh *var. argillosus* M. E. Jones).
Figwort Family (*Scrophulariaceae*).
It was formerly known as Whited's penstemon (*Penstemon whitedii* Piper *subsp. dayanus* [How.] Keck). This rare plant of central Oregon is found only in the drainages of the Deschutes and John Day rivers. Although limited in abundance throughout its range it is currently stable.

Penstemon eriantherus var. argillosus grows to sixteen inches tall. It has a silvery appearance due to the thick, short, whitish, glandular hairs that covers most of the foliage. The basal leaves have stems; the leaves on the stem of the plant are sessile, linear to lance-shaped. The corolla is glandular-pubescent externally, inconspicuously bilabiate with nearly equal, rounded lobes, abruptly expanding as it extends beyond the calyx. It is deep reddish-purple. The long sterile filament is sparsely bearded or glabrous. It can be found in bloom in June and July.

228 229

228

Mountain pride

(*Penstemon newberryi* Gray *ssp. berryi* [Eastw.] Keck).

Figwort Family (*Scrophulariaceae*).

This California penstemon reaches the northern edge of its range in the Siskiyou Mountains in Josephine County, Oregon. It is found in rocky places at fairly high elevations. It has been reviewed for rarity in Oregon, though never actually listed.

Penstemon newberryi ssp. berryi is a shrubby type penstemon arising from woody stems, growing to twelve inches tall. It is mat-forming and its foliage is mostly glabrous. The leaves are elliptic with small teeth on the margins, and are over an inch in length. The corolla is a bright rose-red to reddish-purple, one to one and a quarter inches long, somewhat bilabiate with spreading lobes. The lower lip is densely hairy inside as are the anthers. It blooms in June and July.

229

Peck's penstemon

(*Penstemon peckii* Penn.).

Figwort Family (*Scrophulariaceae*).

This rare species is found only in the central Oregon Cascades in the vicinity of Black Butte and the Metolius River. It grows on flat or sloping ground among ponderosa pines and open grassy areas. It is very limited in range and abundance, but its populations are presently stable.

Penstemon peckii grows up to two feet tall, and is glabrous except for the flowering parts. The rather long, narrow, lance-shaped leaves are in pairs on the stem. The calyx has pointed lobes about one-eighth of an inch long, which are very glandular-puberulent. The corolla varies from a bluish-purple to a pinkish-white in color and it, too, is very glandular-puberulent. It blooms from June to early August.

230 231

230
Short-lobed penstemon;
Narrow-leaved penstemon
(*Penstemon seorsus* [Nels.] Keck.).
Figwort Family (*Scrophulariaceae*).
This rare penstemon of central and southeast Oregon, Crook to Malheur counties, is also found in Owyhee County, Idaho. It grows on stony ridges in dry areas.

Penstemon seorsus reaches twelve inches in height. It has many narrow, linear stem leaves that are paired opposite each other, about one and one-half inches long, with margins that are rolled inward. The calyx consists of pointed lobes; the corolla extends from the calyx without noticeable expansion. The flowers are purplish-blue, covered with very short hairs externally, as is the calyx and the rest of the foliage. The petals are glabrous interiorly. The sterile stamen is bearded nearly its full length with fine yellowish hairs. It blooms in May and June.

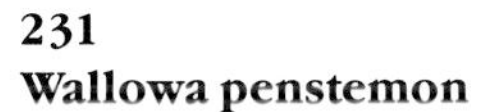

231
Wallowa penstemon
(*Penstemon spatulatus* Penn.).
Figwort Family (*Scrophulariaceae*).
This subalpine to alpine species grows within or above coniferous forests, usually on open, stony slopes at elevations of 5500-8000 feet. It is endemic to the Wallowa and Blue Mountains of eastern Oregon, apparently not occurring in adjacent states of Idaho and Washington. Though common in its habitat at high elevations it is believed to need continued monitoring.

Penstemon spatulatus is a perennial, forming low-growing mats of leaves with flower stems up to ten inches tall. The glabrous leaves, about one to two and one-half inches long, are spatular-shaped with rounded tips, and have entire margins. The inflorescence has one to four dense clusters, and is quite glandular-pubescent. The calyx, about one-eighth inch long, has scarious margins. The blue-violet corolla, about one-half inch long, is glandular and hairy on the outside, has guide lines inside and a bearded palate. The sterile filament is densely bearded at the tip with yellow hairs. It blooms June through August.

232
Silvery phacelia
(*Phacelia argentea* Nels. & Macbr.).
Waterleaf Family (*Hydrophyllaceae*).
This very rare species is found along the Pacific coast from Curry County in southwest Oregon to Del Norte County in California. It grows on sandy bluffs and in dunes close to the ocean. It is considered a narrow, regional endemic which is rare and endangered.

Phacelia argentea is a perennial with many stems which may be up to twenty inches long, some ascending, some decumbent on the sand. The foliage has a silvery sheen produced by many, short, appressed whitish hairs. The leaves are orbicular, thick, not toothed but sometimes with a pair of small lateral lobes near the base. The flowering head is globose; the corolla is white, somewhat longer than the hairy calyx, with very exserted stamens. It can be found blooming from late May to early August.

233
Rock phacelia or Serpentine phacelia
(*Phacelia corymbosa* Jeps.).
Waterleaf Family (*Hydrophyllaceae*).
This rare species is found only on serpentine soil in Curry and Josephine counties in Oregon, and south into northern California.

Phacelia corymbosa is a perennial, has a woody base, and grows to a height of sixteen inches. It is covered with sticky, stiff, whitish, somewhat appressed hairs. Its flowers are white, and form in dense globose heads. The stamens are well exserted; the filaments have long white hairs. The pistil is nearly as long as the stamens. The bristly hairy leaves are lanceolate, entire, sometimes with one or two pairs of small lateral lobes near their base. It blooms from May to August.

234
Mackenzie's yellow phacelia
(*Phacelia lutea* [Hook. & Arn.] J. T. Howell *var. mackenzieorum* Grimes & Packard).
Waterleaf Family (*Hydrophyllaceae*).
This very narrow endemic is found only on pumice hills in eastern Malheur County, Oregon. It is threatened throughout its range.

Phacelia lutea var. mackenzieorum is a rather frail, small-growing plant, reaching eight to ten inches in height. The leaves are small and spatular-shaped. The entire plant, except for the corolla, is covered with whitish hairs. The flowers form in clusters, chiefly terminal on the stem. They are campanulate, bright yellow with brownish veins on the inner surface of the petals, about one-half to three-quarters of an inch long. The sepals are one-half the length of the petals, linear shaped, and rounded at the tip. The stamens do not extrude beyond the throat of the corolla. It may be found blooming in late April and May.

235
Spring phacelia
(*Phacelia verna* How.).
Waterleaf Family (*Hydrophyllaceae*).
Found in Josephine, Coos, Douglas, and Lane counties in Oregon, it grows on moist banks and crevices in basaltic rock mostly in the Umpqua River Valley. It is currently stable but could become threatened in the foreseeable future.

Phacelia verna reaches ten inches in height, is hairy and glandular throughout, and has erect simple or branching stems. It has rather narrow ovate, sessile or very short-stemmed leaves which have entire margins or may be lightly toothed. The scorpioid flower racemes are terminal or may arise from leaf axils. The bell-shaped flowers are white to pale blue, cleft to the middle, and about one-fourth of an inch long. The stamens are slightly longer than the corolla. It is in bloom from April to June.

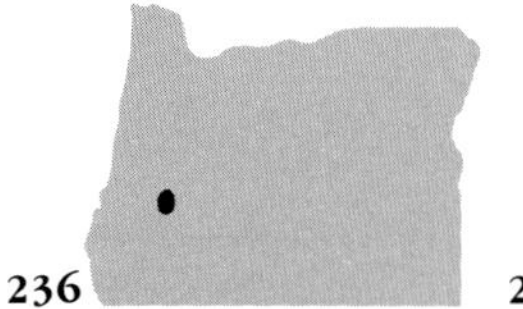

Close-up of flowers

236
Rough allocarya
(*Plagiobothrys hirtus* [Greene] Johnst.).
Borage Family (*Boraginaceae*).
Considered a very narrow endemic, this plant is known to exist in only a few small sites in Douglas County, Oregon. It is endangered throughout its range. A low elevation plant, this one prefers a wet, marshy type grassland that may become dry later in the summer.

Plagiobothrys hirtus is an annual with ascending or reclining stems up to twenty-eight inches long, with linear leaves, paired below, alternate above. The lower part of the plant may be nearly glabrous while the upper part is strongly hirsute with spreading white or yellowish hairs. The racemes are curved, scorpion-like, with the flowers forming along the upper side of the arch. The individual flower stems are very short. The sepals are narrowly lanceolate, very hirsute, and about one-quarter of an inch long. The petals are about twice the length of the sepals, five in number, are rounded at the tip and glabrous. The flowers are pure white, or have a splash of bright yellow in the center. It blooms from mid-June into early July.

237
Fimbriate pinesap
(*Pleuricospora fimbriolata* Gray).
Heath Family (*Ericaceae*) or Indian Pipe Family (*Monotropaceae*).
An herb placed by some in a separate family from *Ericaceae* as it is believed to be saprophytic (living on dead organic material in the soil). It is also said to be mycotrophic (obtaining food by association with a fungus). This plant grows sporadically in coniferous forests in the Siskiyou, Cascade, and Coast Range mountains in Oregon and north through Washington to British Columbia, and south into central California. It is declining in numbers.

The herbage of *Pleuricospora fimbriolata* is white with brownish areas and does not contain chlorophyll. It grows from four to ten inches tall and has a stout stem with scale-like, thick ovate-lanceolate, brown-tipped leaves. The lower leaves are entire or erose, the upper ones are fimbriate, having a margin of coarse, long hairs, giving it its specific name. The flowers, erect on the upper part of the plant, are about one-third of an inch long, with ovate sepals and narrowly elliptic petals, both more or less fimbriate and about one-half inch long. Both sepals and petals are greatly variable. The fruits are white, turning a dark blue at maturity. It flowers from June to August. (The photographed plant is young, as yet without flowers).

238
Shasta fern or Lemmon's shield fern
(*Polystichum lemmonii* Underw.), or (*Polystichum mohrioides* [Bory] Presl. *var. lemmonii* [Underw.]).
This fern grows in Oregon only on serpentine rock, at widely disjunct locations high in the Siskiyou Mountains of Jackson County, where it is considered rare and threatened, but it is more common in California and in Washington in the Wenatchee area, and north to Alaska.

The fronds of *Polystichum lemmonii* are from four to sixteen inches tall, are bipinnate and densely tufted. The new fronds grow among those of the previous year. The leaf stalks tend to become yellowish, as do the scales at the base. The primary divisions of the leaf (pinnae) are ovate, themselves pinnatifid with ovate leaf segments that are overlapping near the tip. The round clusters of sporangia are on the upper pinnae. The indusium, or scale-like covering of the young spores, is dark tinged when ripe.

239
Globose sticky cinquefoil
(*Potentilla glandulosa* Lindl. *ssp. globosa* Keck).
Rose Family (*Rosaceae*).
This plant of the Siskiyou Mountains of Jackson and Josephine counties in southwest Oregon, and Humboldt County in California grows in pine and fir forests at elevations of 4500-7000 feet. Being an extremely variable subspecies, considerable taxonomic problems have arisen. It may in reality be much more common than once was believed.

The stems of *Potentilla glandulosa ssp. globosa* are from eight to sixteen inches tall, and the five to nine leaflets are densely pubescent, and only slightly glandular. The inflorescence is few flowered. Each flower head is globose, giving it its subspecies name. The calyx lobes are about one-quarter inch long, about the same as the petals. The erect petals are creamy or nearly white. It blooms from May to July.

240
Wallowa primrose
(*Primula cusickiana* Gray).
Primrose Family (*Primulaceae*).
The only primrose native to Oregon it is so attractive that it is avidly sought by collectors for use in rock gardens. Unfortunately it is very difficult to establish and usually dies when in a different environment. It is endemic to the Wallowa Mountains in Oregon, and across Hells Canyon in Idaho. It is considered rare and threatened in Oregon. It grows on moist, rocky slopes at about 5500 to 6000 feet, almost always under a ponderosa pine.

Primula cusickiana is perennial, mostly glabrous, from two to eight inches tall. It has oblong-spatulate basal leaves up to three inches long, that are mostly smooth around the edges. There are two to six flowers at the top of a stem, each one about five-eighths of an inch across, with five deep purple or bluish-violet rounded petals, notched at the tip, brilliant yellow at the base extending down into the throat. The anthers and stigma do not extrude from the throat. It blooms soon after the snow melts from April to June.

241
Dalles Mountain or Obscure buttercup
(*Ranunculus reconditis* Nels. & Macbr.).
Buttercup Family (*Ranunculaceae*).
It grows at elevations of 3000-4000 feet in moist areas on sagebrush hills or with juniper or ponderosa pine. This very local and endangered endemic is known only from The Dalles Mountain in Washington and two sites in Wasco County, Oregon. Until recently it was thought to be extinct in Oregon, being last collected in 1895.

Ranunculus reconditis is distinguished from other buttercups by its leaves which are triternately divided, each leaf divided into three segments, each segment again divided into three. Because of this unique leaf it has also been know as *Ranunculus triternatus.* This perennial, which grows to six inches tall, is quite glabrous and has bright yellow petals five-eighths of an inch long. At the base of each petal is a v-shaped yellow gland. It starts blooming in late March and may continue into May.

242
Holly-leaved buckthorn or Red-berried buckthorn
(*Rhamnus crocea* Nutt. *ssp. ilicifolia* [Kell.] C. B. Wolf).
Buckthorn Family (*Rhamnaceae*).
This California plant was not known to exist in Oregon until recently when a small group of them were discovered on private land a short distance east of Ashland in Jackson County. It is considered to be endangered in Oregon but more common in California.

Rhamnus crocea ssp. ilicifolia grows into a large shrub or small tree. It is glabrous, without spines, and is characterized by its leaves which are deeply toothed like a holly leaf. The rather thick, tough, leather-like leaves are from three-quarters to one and one-half inches long. The flowers are unisexual, having either a pistil, or stamens, but not both in the same flower. There are no petals; the petal-like sepals are yellowish-green (reddish on the outer surface), four in number, lance-shaped, sharp-pointed, and about one-quarter of an inch long. The mature fruits are red, sweet, and obovate. It blooms May and June.

Manner of growth

243
Yellow rattle
(*Rhinanthus crista-galli* L.).
Figwort Family (*Scrophulariaceae*).
This circumboreal plant of the north reaches the southern edge of its range in Oregon in Clatsop and Tillamook counties where it is considered rare, though stable in its populations. It is found growing in moist open areas along the coast and on such peaks as Saddle Mountain and Mount Hebo in the northern Oregon Coast Range.

Rhinanthus crista-galli has an erect stem, eight to twenty inches tall, with opposite leaves up to the inflorescence at the top. The sessile, coarsely serrated leaves, as long as one and one-half inches, may be glabrous to villous with very short, white hairs. There may be as many as a dozen flowers in various degrees of bloom in the inflorescence, plus some in the upper leaf axils. Each flower is supported by a broad, leaf-like bract. The corolla is about three-quarters of an inch long, with two lips—the upper lip is a helmet-shaped hood or galea with lateral appendages, the lower lip is three-lobed. The flowers are bright yellow, shading to a rust color with age, and are closely enshrouded by a light green, veined, inflated calyx, about one-half inch long. The lower lip is sometimes spotted with purple. It blooms from June to early August.

244 245

Close-up of flowers and spines

244
Umatilla gooseberry
(*Ribes cognatum* Greene).
Gooseberry Family (*Grossulariaceae*).
Growing on dry hillsides and along the banks of the Umatilla River and its tributaries in northeastern Oregon, this somewhat rare gooseberry may also be found in southeastern Washington and central Idaho.

Ribes cognatum is a shrub with slender stems reaching twelve feet in height. Many branched, the stems have few to many short, slender, brownish prickles and three long stiff spines at the base of each leaf stem. The three-lobed leaves have rounded teeth at the apex. The foliage is covered with fine, glandular hairs. The flowering stems support two to five nodding flowers, white to slightly pinkish. The calyx tube is cylindric, less than one-quarter of an inch long with spreading lobes about the same length. The petals, not spreading, are about two-thirds the length of the calyx lobes, and enclose the cleft style and stamens. The berries are black and smooth. It blooms from April to June.

245
Applegate gooseberry
(*Ribes marshallii* Greene).
Gooseberry Family (*Grossulariaceae*).
This rare wild gooseberry is restricted in Oregon to the Bolan Lake and Grayback Mountain areas of southern Josephine and Jackson counties. It grows in open woods and thickets.

Ribes marshallii grows three to six feet high. The spines are only at the leaf nodes. The leaves are thin, three-lobed, each lobe cleft at the apex, and generally glabrous. The flower stems each support just one flower. The calyx is purplish-red on the inside, greenish and villous on the outside. The calyx tube is short, the lobes, flaring or reflexed, are about one-half inch long. The yellow petals are half the length of the calyx lobes. The stamens are exserted to twice the length of the petals. The berries are dark purple, one-half inch in length, covered with short, sharp spines. It blooms in June and July.

246 247

246
Tracy's mistmaiden or Sea-cliff romanzoffia
(*Romanzoffia tracyi* Jeps.).
Waterleaf Family (*Hydrophyllaceae*).
This small plant grows only on rocky cliffs along the coast from Lincoln and Lane counties in Oregon, to Washington and California.

Up to four inches tall, *Romanzoffia tracyi* has succulent, shiny, softly hairy, kidney-shaped leaves, two and one-half inches wide, with seven to nine notches on the edges. Each raceme consists of several tightly arranged flowers and buds. The flowers are white, veined with green, yellowish in the throat. The corolla is funnelform, about one-third of an inch long. The five stamens have white anthers and are not exserted from the corolla tube. It blooms in April.

247
Wapato
(*Sagittaria latifolia* Willd.).
Water-plantain Family (*Alismaceae*).
Also called Broad-leaved arrowhead because of its wide, arrow-shaped leaves, and Indian potato, because its tuber was an important food to the Indians. Once very common throughout the marshy areas of the lower Willamette and adjacent Columbia rivers, it nearly disappeared from this area due to the destruction of its habitat. Where protected, it is now coming back. It may be found west of the Cascades from Vancouver Island to California, and is more abundant in central and eastern United States.

Sagittaria latifolia is monocotyledonous, grows in shallow water or mud to a height of from one to four feet, is a perennial, has arrow-shaped, parallel veined, sagittate leaves with the basal lobes strongly divergent backwards, sometimes longer than the rest of the leaf blade. The flowers are usually in whorls of three. They are white with three round petals, each about one-half inch long. The flowers may be staminate or pistillate, or both. The staminate are usually above the pistillate on the flowering stem. The anthers are bright yellow. It blooms from mid to late summer.

248 249

248
Arctic willow
(*Salix arctica* Pall.).
Willow Family (*Salicaceae*).
Much more common to the north of Oregon, this willow of the Arctic is found here only on the high peaks of the Wallowa Mountains and on Steens Mountain. It is rare and possibly threatened in Oregon.

Salix arctica is a low growing, matted willow with creeping stems. The branches are glabrous, brownish in color, seldom more than three to four inches above the ground, often running just under the ground. The leaves, elliptic in shape, are one to two inches long. The catkins are unisexual, with the male (staminate) and the female (pistillate) catkins occurring on separate plants. The female catkin is quite large, sometimes over two inches long. In this species, the catkins and the leaves develop at the same time. The scales that enclose the flowers in their early stage are dark brown and hairy. It produces catkins after snow-melt, usually mid-summer.

249
Cascade willow
(*Salix cascadensis* Cockerell).
Willow Family (*Salicaceae*).
This low-growing alpine willow has been reported from summits and in moist alpine meadows in the Wallowa and Ochoco Mountains in Oregon, north into Washington, and British Columbia, and east to Montana, Wyoming, Colorado, and Utah. It is rare in Oregon and needs to be monitored.

Salix cascadensis is only about five inches tall, prostrate, creeping, and mat-forming. The leaves are small, to five-eighths of an inch in length, narrow, tapered to both ends, entire, shiny green on the upper surface, light green underneath, glabrous at maturity, and quite thick with a prominent mid-rib. The catkins which appear at the same time as the leaves, are less than an inch in length. It produces catkins in July.

250  251

250
Drummond's willow

(*Salix drummondiana* Barrat ex Hook.).
Willow Family (*Salicaceae*).
This willow, widespread west of the Rocky Mountains from Alberta to New Mexico, is found in Oregon on Steens Mountain in the vicinity of Fish Lake, Jackman Park, and Whorehouse Meadow, and possibly a few sites in the Wallowas and Blue Mountains. It grows in flat, marshy meadows and along streams at elevations of about 7500 feet. It is rare in Oregon.

Salix drummondiana, a small tree, may reach twelve feet in height. The leaves are much longer than wide, tapered to both ends, widest toward the apex. Quite variable, they are usually shiny green on the upper surface, with dense white hairs underneath and are from one to three inches long. The male and female catkins are found on separate plants and may be up to one and one-half inches in length. It produces catkins in May at Fish Lake.

Female catkins

Male catkins

251
Peck's snakeroot

(*Sanicula peckiana* Macbr.).
Parsley Family (*Apiaceae*).
This rare Siskiyou endemic is found on dry slopes and open woods in southern Curry and Josephine counties in Oregon and in northern California where it is endangered in parts of its range.

Sanicula peckiana is a perennial herb, with basal leaves, two and one-half inches long on stems as long that are prominently "winged" with a thin expansion bordering each side of the stem. The pinnate leaves, along with their primary segments, are glabrous and sharply and irregularly toothed. The flower stems, generally leafless, are quite stout, grooved longitudinally, and somewhat reddish in color. The yellow flowers are in dense umbels above small leaf-like bracts. It blooms April to May.

252
Snow plant
(*Sarcodes sanguinea* Torr.).
Heath Family (*Ericaceae*).
This very striking red saprophyte is known from southwest Oregon in the Siskiyou, Cascade, and Coast Range mountains south to the mountains of southern California and east to the Sierra Nevadas in Nevada. Its most northern reported site is above the North Umpqua River in Douglas County, Oregon. It grows in coniferous forests at 4000 to 8000 feet elevation, in thick humous soil. Though apparently not threatened in Oregon, neither is it seen in abundance.

Sarcodes sanguinea is a brilliantly colored plant, without chlorophyll, that lives on the dead organic material found in the soil. It grows to twelve inches tall, has scale-like, bright red leaves (one to three inches long) on its stout whitish stem. They are ciliated with white hairs especially along the margins. The numerous urn-shaped flowers have bright red sepals that are lance-shaped, rounded, and have ciliated margins. The sepals closely enclose, but are slightly shorter than the bright red petals. The petals are from one-half to three-quarters of an inch long. The perianth is persistent and becomes distended and globe-like in fruit. It blooms May to July.

253
Strawberry saxifrage
(*Saxifraga fragarioides* Greene).
Saxifrage Family (*Saxifragaceae*).
This uncommon saxifrage can be found growing on dry cliffs in the high Siskiyou Mountains of southern Josephine and Jackson counties in Oregon and south into the Trinity Mountains of northern California.

Saxifraga fragarioides is a perennial herb, reaching ten to twelve inches in height, with chiefly basal leaves. The stem leaves are reduced mostly to bracts. The almost glabrous leaves are wedge-shaped, toothed along the rounded distal edge, and entire along the sides. They have a jointed attachment to their stems and ultimately disarticulate (separate) at that joint, thus being also known as the joint-leaf saxifrage. The inflorescence, densely glandular-pubescent, is in the form of a loose panicle. Each branching flower stem is subtended by a narrow bract. Though the tiny petals are white, the over-all appearance of the flower is yellow due to the bright yellow carpels in the center of the flower. The petals are elliptic to spatulate in form and are attached by a narrow stalk called a claw. It blooms in June and July.

254
Saddle Mountain saxifrage
(*Saxifraga hitchcockiana* Elvander); until recently, it was known as *Saxifraga occidentalis* Wats. *var. latipetiolata* C. L. Hitchc.
Saxifrage Family (*Saxifragaceae*).
It is found near the summits of some of the higher peaks in the north Oregon Coast Range in Clatsop and Tillamook counties, including Saddle Mountain. Because of its limit to the above geographic area it has been listed as an endemic, and has been determined to be threatened throughout its range.

Saxifraga hitchcockiana grows ten to twelve inches tall, has white, five-petalled flowers, each petal about one-eighth of an inch long. It is differentiated from other saxifrages by its rounded basal leaves that are covered with glandular yellow hairs, and that taper gradually to a broad petiole that is much shorter than the leaf blade. Its flowers bloom from late June to July.

255
Purple saxifrage
(*Saxifraga oppositifolia* L.).
Saxifrage Family (*Saxifragaceae*).
This circumboreal species of high mountain peaks reaches its southern limit in the Rocky Mountains of Montana, the Cascade and Olympic mountains of Washington, and the Wallowas and Blue Mountains of northeastern Oregon. Though its range is small in Oregon, it is quite abundant in its high alpine habitat where it can be found growing on solid rock. It is more common further north.

Saxifraga oppositifolia forms a beautiful spreading mat of small, closely arranged, geometric leaf rosettes. The leaves are coarsely ciliate, ovate, about one-quarter of an inch long. They are thick, tough, tenacious to the rock, and firm with stiff, sharply pointed tips. They are in the form of opposite pairs concentric within each other, consequently the species name "*oppositifolia*". The flowers, only two inches tall, consist of five obovate, deep rose-colored petals, purple-veined, surrounding ten stamens and a double-styled pistil leading to a superior ovary. It blooms in July.

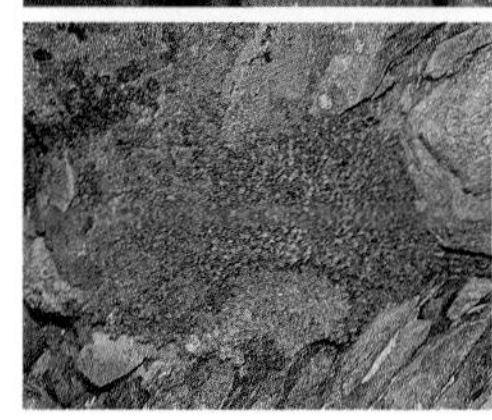

Habitat view

256
American scheuchzeria
(*Scheuchzeria palustris* L. *var. americana* Fern.).
Arrow-grass Family (*Juncaginaceae*).
This widespread but very rare aquatic grows in ponds and along streams in the Oregon Cascades, north into Alaska, south to California, across North America and in the eastern hemisphere. It is threatened in Oregon and was presumed to be extinct in California until rediscovered this year.

Scheuchzeria palustris var. americana is a perennial that grows to twenty inches in height. Its leaves are narrow, and sheathed loosely at the base. Stem leaves are much reduced and bract-like. The flowers are few in a rather tight raceme; the terminal flowers bloom last. The perianth is whitish with segments one-eighth of an inch long. It blooms in June and July.

257
Oregon fetid adder's-tongue
(*Scoliopus hallii* Wats.).
Lily Family (*Liliaceae*).
This small lily may be found growing along streams in moist woodlands in the western Cascades and the Coast Range from Tillamook County south to the Siskiyous. Though not considered as a threatened species in Oregon, it is rarely seen.

The stem of *Scoliopus hallii* is short, mostly underground, and is sheathed by the older leaf bases. The two leaves, unmottled or very lightly mottled green, are elliptic to oblong in shape, parallel veined, pointed, and about four to six inches long. The flowers, from one to eight on a plant, occur singly at the apex of a slender stem about two and one-half inches tall. The perianth consists of three recurved sepals, one-third of an inch long, which are greenish-white with dark purple veins, and three erect petals, about as long, but narrower, and arched slightly inward around the three anthers and the triple-divided, recurved stigma. The fruits are capsules about one-half inch long. As the plant goes into fruit, the flower stems weaken, allowing the capsule to lie on the ground. It blooms in February and March.

258 259

258
Heckner's stonecrop

(*Sedum laxum* [Britt.] Berger *ssp. heckneri* [Peck] Clausen).
Stonecrop Family (*Crassulaceae*).
Found in Curry, Douglas, Jackson and Josephine counties in southwest Oregon and in adjacent California, it grows in dry areas in peridotite and gabbro rock outcrops at elevations of 4500 to 5300 feet. The photographed plant was seen near the summit of Fiddler Mountain. This siskiyou endemic is presently rare but stable throughout its narrow range but needs to be monitored.

The basal leaves of *Sedum laxum ssp. heckneri*, about an inch long, form a succulent rosette, are flat, thick, broadest near the tip, and are glaucous with a whitish bloom. The stem leaves are similar. The pink flowers form clusters at the top of the four to eight inch reddish stems. The plant is without hairs. It blooms in June and July.

Close-up of flowers

259
Glandular stonecrop

(*Sedum moranii* R. T. Clawsen; also known as *Sedum glanduliferum* [Hend.] Peck, and *Gormania glandulifera* [Henderson] Abrams).
This very rare stonecrop is found only along the Rogue River in the vicinity of Galice where it is threatened throughout its range. It grows on serpentine cliffs and outcrops at a narrow elevational range, 600 to 900 feet.

Sedum moranii is distinguished by its deep red to purplish-red stems and foliage and bright greenish-yellow flowers. The inflorescence, including the petals, is densely glandular-pubescent inside and out; the leaves and stem are quite glabrous. The plant is a perennial, grows erectly six to eleven inches tall, has succulent, spatulate basal leaves, and similar, though smaller alternating stem leaves. The petals are one-half inch long. The stamens are about one-quarter inch long, slightly longer than the calyx. It blooms from May to mid-June.

260
Sierra sedum

(*Sedum obtusatum* A. Gray *ssp. retusum* [Rose] Clausen).
Stonecrop Family (*Crassulaceae*).
This is a California plant that extends north into southern Jackson County, Oregon. It has just recently been determined to be more plentiful in this state than once believed. It grows on rocky slopes and ridges.

Sedum obtusatum ssp. retusum is two to six inches tall. The stem is stout; the pale green, somewhat glaucous, flat, fleshy, spatulate basal leaves, nearly an inch in length, are rounded at the apex and stand nearly erect in cup-like clusters. The corollas, one-quarter to one-third of an inch in length, are pale yellow to slightly pinkish. It blooms in June and July.

Close-up of flowerheads

261
Ertter's senecio

(*Senecio ertterae* Barkley).
Composite Family (*Asteraceae*).
This newly-discovered species, described in 1978, is a local endemic found growing only in volcanic tuff at about 4000 feet elevation in the vicinity of Leslie Gulch in eastern Malheur County, Oregon. It is a candidate for federal listing as endangered.

Senecio ertterae is an annual herb about 24 inches tall. Its foliage is succulent, grayish-green with silvery, wool-like, matted hairs. The leaves grow from the stem, are long toothed, and up to three inches long, becoming smaller toward the top. There may be as many as six to ten flower heads in the inflorescence. The eight to thirteen ray flowers in each head are yellow, and less than one-quarter inch long; the fifty disk flowers are also bright yellow. It blooms from June through August.

262
Siskiyou butterweed
(*Senecio hesperius* Greene).
Composite Family (*Asteraceae*).
This candidate for federal endangered status is found on serpentine soil in the Illinois River valley in southern Josephine County. It grows in open forests of Jeffrey Pine.

Senecio hesperius is a perennial herb about twelve inches in height. It has wavy-margined, oblong, nearly entire basal leaves, one inch long, and only two to three smaller bract-like stem leaves. Mostly glabrous, there may be a few white, woolly hairs at the leaf attachments and around the flowering head. There are from one to five flower heads per plant. The six to ten ray flowers are bright yellow and less than one-half inch long. The disk flowers are also bright yellow. The blooming time is from April to June.

263
Verrucose sea-purslane
(*Sesuvium verrucosum* Raf.).
Carpet-weed Family (*Aizoaceae*).
This wide-ranging species of California, Nevada, Texas, Utah, Kansas, northern Mexico, and reportedly southern Brazil, has only recently (1977 and 1979) been discovered in Oregon from sites in the Warner Valley of Lake County, and Tum Tum Lake in Harney County. It grows in low, more or less salty areas near the muddy edge of ponds, lakes, and sinks. It is threatened in Oregon but more common elsewhere.

Sesuvium verrucosum, a prostrate or decumbent fleshy, perennial herb, has stems to twenty inches long with opposite, succulent, spatulate, entire leaves, about one and one-half inches long, which are covered with minute papillae. The stemless, showy flowers are star-shaped, with five pointed sepals that are bright pink on the inside, greenish on the back, and about one-quarter inch long. There are no petals. There are numerous stamens and three to five styles. It blooms from late June to August in our area.

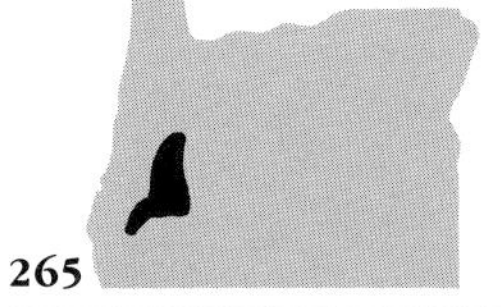

264
Meadow sidalcea
(*Sidalcea campestris* Greene).
Mallow Family (*Malvaceae*).
This once very common wildflower in native grasslands of the Willamette Valley is now limited to fence-rows, roadsides and ditch-banks in the north and western parts of the valley, where many are lost each year to roadside spraying to which they are very susceptible. Although still found in areas around Salem, its populations need protection.

Sidalcea campestris is a stout perennial herb growing to five feet tall. It is somewhat pubescent, has large fan-shaped basal leaves that are seven- to nine-lobed and smaller stem leaves that are deeply divided into five to seven narrow, pointed lobes. The flowering raceme, up to twelve inches long, is loosely arranged and has numerous pale pink to nearly white flowers with petals one-half to one inch in length, broadest and notched at their periphery. Some plants may be pistillate only, others may have stamens, too. The strictly pistillate flowers appear quite differently, being smaller and deeper in color. This is true in most of the sidalcea species. It blooms from April to early July.

265
Cusick's sidalcea
(*Sidalcea cusickii* Piper).
Mallow Family (*Malvaceae*).
This rare species, found in the southern Willamette Valley, and the Umpqua and Coquille river valleys, is presently stable, but should be monitored.

Sidalcea cusickii is a perennial growing to five feet tall, with a stout, hollow stem supporting one to several long, densely flowered, spike-like racemes. The petals, over one-half of an inch long, are a rose-pink aging to a deeper purple, prominently veined, and broadest and notched at the tip. Some flowers are perfect with both stamens and pistils; others just have pistils, and are generally smaller. Pistils, and stamens when present, form a column in the center of the flower, spreading at the apex. The calyx lobes, broadened above the base, differentiates it from another species, *Sidalcea virgata*, which appears very similar and is more common in western Oregon. The base leaves are deeply cleft into seven to nine lobes, are four to six inches across. The stem leaves may be cleft or toothed, the upper ones becoming linear and bract-like. It blooms from May to July.

266
Bluff mallow or Bristly-stemmed sidalcea
(*Sidalcea hirtipes* C. L. Hitchc.).
Mallow Family (*Malvaceae*).
This showy plant grows on grassy bluffs above the sea from northern Lincoln County, Oregon to southern Washington, and on Saddle Mountain in the Oregon Coast Range. It is considered to be rare in Oregon, threatened in Washington.

Sidalcea hirtipes grows two to three feet tall. The stem and sepals are more or less stiff-hairy; the rest of the plant is quite glabrous. Its basal leaves, two and one-half inches across, are not deeply divided but are generally notched around the edges. They have hairy stems like the main stem even though the leaves are smooth. The flowers, clustered at the top of the stems, are a rich rose-purple in color; the etals are three-quarters of an inch long. The stamens and pistil form a central column. It blooms from June to August.

267
Checker bloom
(*Sidalcea malvaeflora* [DC] Gray ex Benth. *ssp. elegans* [Greene] C.L. Hitchc.).
Mallow Family (*Malvaceae*).
This rare subspecies of a highly variable species is found in stony areas in the vicinity of Rough & Ready Botanical area and along Oregon Mountain Road in southern Josephine County, south into Del Norte and Siskiyou counties, California. It is rare throughout its range.

Sidalcea malvaeflora ssp. elegans is a perennial growing to two feet in height. It has relatively few flowers on very short stems near the top of the main stem. The flowers are rosy-pink, usually white-lined, and nearly an inch long. The stem is sparsely hairy and glaucous below, glabrous and very slender above. The leaves are a shiny deep green, glabrous, round in outline, divided almost to the center into five lobes, some of which are also deeply divided. The stem leaves are few and small. The bracts under the flower stem are linear and often two-cleft. It blooms in May and June.

268
Nelson's wild-hollyhock
(*Sidalcea nelsoniana* Piper).
Mallow Family (*Malvaceae*).
This very rare plant, endemic to the Willamette Valley and adjacent Coast Range has become famous in recent years in the part it has played in delaying the construction of a dam in the Walker Creek wet-lands above McMinnville. *Sidalcea nelsoniana* is a candidate for the federal status of endangered species.

Sidalcea nelsoniana is differentiated from other *Sidalcea* by its small, rich rose-pink flowers, with petals only one-quarter to one-half of an inch long, by its almost glabrous calyx, and by the simple pubescence on the stem. It grows to three feet tall, and has orbicular green leaves; the basal leaves are divided into seven rather shallow rounded lobes. The upper leaves are divided more deeply. This plant is highly susceptible to herbicide sprays; it was once wide-spread in the Willamette Valley. It blooms in June and July.

269
Oak Flat sidalcea
(*Sidalcea setosa* C. L. Hitchc. *ssp. querceta* C. L. Hitchc.).
Mallow Family (*Malvaceae*).
This very rare taxon was once known from one small area high on a bluff above the Illinois River, near Agness, at the edge of a field called Oak Flat. It is now believed to be extinct, as the plant now found at its only known site has been determined to be *Sidalcea malvaeflora ssp. elegans*, as shown in the accompanying photograph. *Sidalcea setosa ssp. querceta* was very susceptible to grazing.

While some botanists feel this was not a definite, separate subspecies from *Sidalcea setosa ssp. setosa* found in Josephine County, the plants at Oak Flat were quite uniform in nature, having some constant differences. *Ssp. querceta* had few flowers and little or no rootstock by which it was spread, while *ssp. setosa*, with many, densely arranged flowers, does have a short rootstock. *Ssp. querceta* had a smaller calyx by one third, with conspicuously finer star-shaped hairs, was not as bristly, and lacked the long hairs common to *ssp. setosa*. The flowers were rosy pink, and it bloomed in May and June.

270
Bristly sidalcea
(*Sidalcea setosa* C. L. Hitchcock *ssp. setosa*).
Mallow Family (*Malvaceae*).
This rare species of woodlands and meadows is endemic to the Siskiyou Mountains of southern Josephine County in Oregon and Siskiyou County, California.

Sidalcea setosa ssp. setosa grows four to five feet tall from a short, spreading rootstock. It has base leaves that are fan-shaped, up to six inches across, shallowly lobed into five to nine segments; the stem leaves are smaller and more deeply divided into narrow linear lobes. The calyx which is one-quarter of an inch long, and very bristly, enlarges after flowering. The flowers, arranged in a dense spike-like raceme up to three inches long, are light to deep pink with petals from one-quarter to one-half inch in length. The upper part of the plant, including both sides of the leaves, is covered with minute, star-shaped hairs; the lower part has fine, soft hairs. It blooms in June and July.

271
Cascade Head catchfly
(*Silene douglasii* Hook, *var. oraria* [Peck]; also known as *Silene oraria* Peck).
Pink Family (*Caryophyllaceae*).
This rare and endangered plant, found only on three headlands along the Oregon coast in Tillamook County, grows on steep bluffs and grassy slopes facing the ocean.

Silene douglasii var. oraria is generally a small plant, about six to ten inches tall, with slender, branching stems, covered with a very short, fine, matted, woolly-like hair. The sessile, acuminate leaves are paired, and linear-lanceolate, one to eight pairs on the stem. Most of the leaves, however, are matted at the base. The calyx, nearly one-half inch long, is tubular and inflated. The flowers are white, tinged with pink or purple. The petals are one-half to three-quarters inch long, each divided into two terminal lobes, and each with two appendages near the base. Much study is presently being done with these plants to determine their stability, and why they are where they are, but nowhere else. It blooms from April to July.

272
Bolander's catchfly
(*Silene hookeri* Nutt. *ssp. bolanderi* [Gray] Abrams).
Pink Family (*Caryophyllaceae*).
This rare and endangered plant of serpentine slopes once found in Josephine County in Oregon, is now believed to be extinct in this state. It still exists in California where it is quite common.

Silene hookeri ssp. bolanderi has upright stems two to six inches tall. The leaves are one and one-half to three inches long, and about one-half inch wide. The calyx is tubular, one-half to three-quarters inch long. The stems, leaves, and calyx are all covered with numerous short, white hairs. The flowers, in clusters of one to nine, are generally white but may be pinkish or lightly violet. The petals, an inch long, are deeply divided into narrow segments that are equal in width. It blooms May and June.

273
Hooker's pink or Dusty pink
(*Silene hookeri* Nutt. *ssp. pulverulenta* [Peck] Hitch. and Maguire; also *Silene pulverulenta* Peck).
Pink Family (*Caryophyllaceae*).
This rare plant of the Siskiyou Mountains of eastern Josephine and western Jackson counties in south-west Oregon has recently been discovered in Del Norte County, California.

Silene hookeri ssp. pulverulenta grows about five inches high in dry, sandy soil. Its pubescence is extremely viscid. The lower leaves are large, up to three inches long and one and one-half inches wide; the upper ones are much smaller. The calyx is tubular, somewhat inflated, about five-eighths of an inch long. The flowers vary from pink to almost white, are numerous, and have five petals, each nearly an inch in length, each deeply divided into four lobes, the two in the middle being the largest. Each petal is attended by two small appendages at the throat. It blooms in May and June.

274 275

274
Suksdorf's campion

(*Silene suksdorfii* Robins).
Pink Family (*Caryophyllaceae*).
This rare plant grows in scree slopes above timberline on the peaks of the Cascade Mountains, from Mt. Theilsen in Oregon to Mt. Baker in Washington. Limited in numbers, it is currently considered stable.

Silene sudsdorfii is about four inches tall with one to three flowers on a stem. Its herbage is glandular, especially at the top. The leaves are very numerous, linear in shape, an inch long, and paired on the flowering stem. The calyx is cylindrical, somewhat inflated, one-half inch long, with conspicuous purple ribs. The petals, two-lobed and only slightly longer than the sepals, are creamy white shading to a light purple. There are small appendages near the throat on each petal which may be white or deep purple. It blooms from July to September.

275
Hitchcock's purple-eyed grass

(*Sisyrinchium hitchcockii* Hend.).
Iris Family (*Iridaceae*).
This newly-described species of the Umpqua and southern Willamette valleys in Oregon is on the Review List in Oregon as there is little information regarding it.

The stems, fourteen to eighteen inches in height, rise singly or in clusters, and are erect or somewhat decumbent. The foliage is generally glabrous and glaucous, though the underside of the tepals and the ovary may be somewhat glandular. The basal leaves are about two-thirds the height of the stem and about one-quarter of an inch wide. There may be several clasping, linear-shaped stem leaves. The two upper-most, nearly equal in length, form a bract-like spathe immediately below the flower head. Often the spathe and some of the leaves are reddish in color. The six tepals, each about one-half of an inch long, are deep purple with darker purple veins, and occasionally yellow at the base. They are oblong, rounded, or indented at the tip, from which extends a sharp point. The filaments are united in a column tipped with bright yellow anthers. It blooms from late May into July.

Close-up of flower

276
Pale blue-eyed grass
(*Sisyrinchium sarmentosum* Sudsd. ex Greene).
Iris Family (*Iridaceae*).
Until very recently this plant was known only in Skamania and Klickitat counties in Washington, where it grows along the margins of wet meadows. It has just been recorded in Oregon in Clackamas County, near Little Crater Lake south of Mt. Hood. It is a threatened species in Washington and its status in Oregon has just recently been determined endangered.

Sisyrinchium sarmentosum grows in small tufts to a height of about eight inches. The leaves, basal, and nearly as tall as the stem, are narrow and grass-like. The stem is leafless, topped with two very narrow bracted spathes, the outer, one and one-half to two inches long; the inner spathe is one to one and a quarter inches long, both longer than the slender pedicels of the one to three exserted flowers. The flowers are pale violet with six perianth segments one-third of an inch long, each having a sharp, slender point at the tip. It blooms June and July.

Minute greenish flowers

Stem with weak prickles

277
Greenbrier
(*Smilax californica* [DC.] Gray).
Lily Family (*Liliaceae*).
This strange plant is threatened in Oregon but is more common in California. It can be found in Oregon along the Rogue and Illinois rivers and south into California. It grows along streambanks and in thickets, often climbing over and covering other plants.

Smilax californica is an unusual lily. It is a climber with tendrils, and has soft thorns. The leaves are heart-shaped, parallel-veined, glabrous, pointed, and four to five inches long, with tendrils at the base of the somewhat flattened petiole. The flowers are very small, dioecious, greenish-white, five to twenty in an umbel stemming from a leaf axil, and produce round black berries one-quarter of an inch in diameter. It blooms from May to June.

278 279

278
Western sophora
(*Sophora leachiana* Peck).
Pea Family (*Fabaceae*).
Western sophora is another rare and unusual plant of southwest Oregon, growing over a peridotite substrate in mixed evergreen-oak woods habitat, often in open areas. It is found only in western Josephine County, and is the only member of its genus in the western states. It is restricted in its dispersibility because of its low proportion of viable seeds.

The greenish-white flowers of *Sophora leachiana* are about one-half of an inch in length, occur in racemes of twenty to fifty at the top of stems twelve to sixteen inches tall. The calyx is enlarged or humped on one side near its base. The configuration of its growth is striking. Arising singly from its rootstock, it has no basal leaves. Instead it puts forth several alternate, long (up to ten inches), arching pinnate stem leaves, with from nineteen to thirty-three leaflets. The leaves become more dense toward the top, are sometimes branched, and are covered with short, matted, woolly hairs giving it a somewhat grayish appearance. It blooms from late April into June.

Close-up of flowers

279
Western ladies' tresses
(*Spiranthes porrifolia* Lindl.).
Orchid Family (*Orchidaceae*).
This rare orchid grows in swampy ground, mostly east of the Cascades, north to Washington where it is described as a sensitive plant, east to Colorado, and south to California.

Spiranthes porrifolia is a glabrous plant with a slender stem four to twenty inches tall. The several leaves arising from near the base are narrow, and up to ten inches long, and become short, sheathing bracts higher on the stem. The greenish-white flowers, usually more than twenty in the two to five inch spike, appear to be in spiralling rows. The sepals are sticky-pubescent. The lip is nearly triangular and is only slightly constricted below the tip. There are two nipple-shaped appendages at the base of the lip. There is no spur. The perianth parts are narrower than on the similar, more common species, *Spiranthes romanzoffiana* Cham. It blooms from June to August.

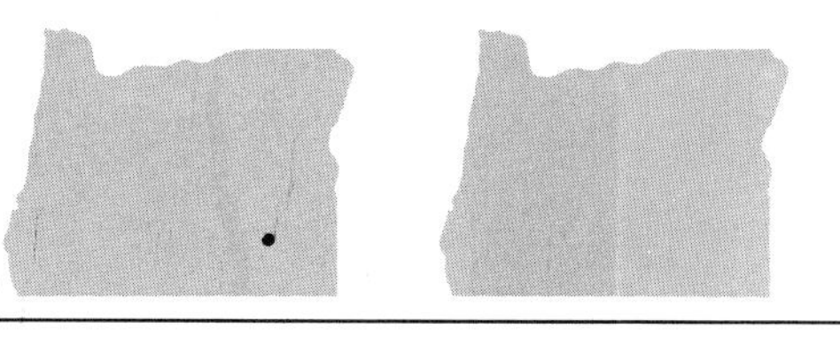

Manner of growth

Close-up of flowers of parent plant

Close-up of flower of Malheur wire-lettuce

280
Malheur wire lettuce or Skeletonweed
(*Stephanomeria malheurensis* Gottlieb).
Composite Family (*Asteraceae*).
This is one of the very few species that is Federally described as endangered in Oregon. Its range is extremely small, a few acres on a hill several miles south of Burns, which is known to be a valid mining claim. It is believed to be one of the very few known examples of a newly evolved annual plant, with its parent plant, *Stephanomeria exigua* Nutt. *var. coronaria* [Greene] Gottlieb being nearby. The parent species is quite common and wide-spread. The new species maintains its distinction by using an inbreeding mode of reproduction, and does not cross with its parent plant which is an outcrosser. The numbers of this plant varies greatly from year to year depending upon the amount of precipitation at the beginning of the growing season, and it does not compete well with other plants. It appears to do better in years after which fire has destroyed many of its competing plants.

Stephanomeria malheurensis is a sparse plant, growing to twenty inches in height, having wiry, branching stems, and leaves that are reduced to bracts. The new species is diploid, having double the basic number of chromosomes. Most of its flowers are white. It has from five to eleven flowers per head, and the seeds are fifty percent larger than those of the parent plant. The pappus bristles are divided and protrude further out of the seed head. (*Stephanomeria exigua var. coronaria* is about 95% pink with only three to five flowers per head, and has smaller seeds, and shorter, undivided pappus bristles.) *Stephanomeria malheurensis* blooms in July and August, its flowers open quite early in the morning, and close when the sun strikes them.

281
Howell's streptanthus
(*Streptanthus howellii* Wats.).
Mustard Family (*Brassicaceae*).
Endemic to the Siskiyou Mountains of Curry and Josephine counties in Oregon and Del Norte County in California, this rare and endangered species grows in brushy areas or open woods in dry serpentine soil.

Streptanthus howellii is a perennial, growing erectly to two feet tall. Its leaves are spatulate, untoothed, without stems but not clasping the main stem. It is without hairs, somewhat glaucous with a powdery green appearance. Its flowers form a loose raceme at the top of the stem. The petals, less than one-half inch long, are dark purple, and only slightly longer than the sepals. The seeds form in-curved siliques that are noticeably flattened and up to three inches long. It blooms from July to September.

282
Violet suksdorfia
(*Suksdorfia violacea* Gray).
Saxifrage Family (*Saxifragaceae*).
This rare plant, which ranges from the Columbia River Gorge to British Columbia, Idaho, and western Montana, was thought to be extinct in Oregon from 1895 until it was rediscovered in 1979 near Mosier in the Columbia Gorge. It was also found that year on the Klickitat River in Washington, its type locality. It is endangered in Oregon. It grows in wet rocky areas at low elevations.

Suksdorfia violacea reaches ten inches in height. The leaves are round or kidney-shaped with three to five shallow lobes. Sometimes a stem-like appendage continues beyond the end of the leaf to form another leaf blade. The basal leaves have long stems; the stems on the cauline leaves are very short. The foliage is glandular-puberulent throughout. The flowers, three-quarters of an inch in diameter, are single, or a few at the top of the stem. The five petals are purple or white, becoming a greenish-white near the throat. They are spatulate in shape, rounded at the tips, and arranged like overlapping spokes in a wheel. The pistil and stamens are contained within the tube of the corolla. It blooms in March and April.

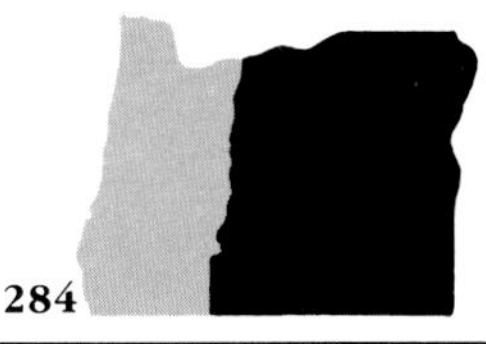

283
Oregon sullivantia
(*Sullivantia oregana* Wats.).
Saxifrage Family (*Saxifragaceae*).
This plant which grows on shaded, perpetually wet, rocky areas, usually within the spray zone of a waterfall, is found only on the lower Willamette River, and in the west end of the Columbia Gorge in Oregon and Washington. It is considered to be a threatened endemic.

Sullivantia oregana has very slender stems that are glandular and puberulent toward their tips, and glabrous toward their base. The glabrous leaves are triangularly cleft and up to two inches across. The calyx is glabrous with sharp, triangular lobes. The five petals are white, about one-quarter of an inch long; the five stamens are tipped with yellow anthers on short filaments located entirely within the tube of the flower. It flowers from May to July.

284
White-stemmed swertia
(*Swertia albicaulis* [Griseb.] Kuntze. *var. idahoensis* [St. John] Hitchc., has also been known as *Frasera albicaulis* Dougl.).
Gentian Family (*Gentianaceae*).
This plant, rarely seen, occurs in dry areas east of the Cascades, extending north into Washington, and east into Idaho where it is periodically being monitored for rarity.

Swertia albicaulis var. idahoensis grows to twenty inches tall. Its foliage is covered by a very fine puberulence, giving it an over-all grayish-green appearance. The base leaves are narrowly spatular in shape, about six inches long, with white margins. The stem leaves are similar, but smaller and are in opposite pairs. The flowers are four-petalled, light blue or lavender, with a greenish, rib-like gland in the center of each petal. The gland is bordered by rather long hairs (setae) that curve inward and interlace over it. The stamens are also lavender. It blooms in May and June.

285 286

285
Mountain kittentails

(*Synthyris missurica* [Raf.] Penn. *ssp. hirsuta* Penn.).
Figwort Family (*Scrophulariaceae*).
The flower pictured is NOT *subspecies hirsuta*, but the more common, similar *ssp. missurica* found on a summit in the Wallowa Mountains. *Ssp. hirsuta* was collected only once, in Douglas County, in 1881, by Thomas Howell. It has not been reported since, and is probably extinct. It differs from the *ssp. missurica* by having larger flowers, and by having brownish hairs on the short flower stems. (Current speculation suggests *Synthyris missurica ssp. hirsuta* may have been misidentified and mislabeled as to locality, and that it could actually be a synonym of *Synthyris stellata* Penn.).

Synthyris missurica ssp. missurica is found in coniferous forests at elevations of 4500 feet to over 9000 feet in several of the mountain ranges of Oregon, Idaho, Washington, and California. It is also known as Lewis and Clark's synthyris. It grows four to sixteen inches tall, has orbicular, sharply-toothed leaves, one to two inches in diameter, that are glabrous and even shiny when young. The raceme of flowers is dense, and is two to seven inches long. The deep bluish-purple flowers have one upper and three smaller lower petals. The two well-exserted stamens have bluish filaments and anthers. The bright green sepals are linear in shape. It blooms from April to June.

Close-up of flowers

286
Fringed synthyris

(*Synthyris schizantha* Piper).
Figwort Family (*Scrophulariaceae*).
This rare plant is known only from a few mountain peaks in the north Coast Range in Oregon, and from the Olympic and Cascade Mountains of Washington.

The flowers of *Synthyris schizantha* are in a terminal raceme, about three inches long. The corolla is purple, striped with darker purple, shading to white and then green in the throat. The four lobes of the corolla, one upper and three lower, are deeply incised giving the plant its descriptive name. The petals are nearly one-half inch long, twice the length of the calyx which also has four lobes. The leaves are kidney-shaped, up to three inches broad, deeply veined, dark purplish-green in color, cupped and irregularly toothed. The stems and the veins of the leaves are villous with long, soft, white hairs. The two stamens, tipped with reddish anthers, and the pistil are long exserted from the corolla. It blooms in May and June.

Close-up of flowers

287 288

Close-up of flowers

287
Columbia synthyris, or Kittentails
(*Synthyris stellata* Penn.).
Figwort Family (*Scrophulariaceae*).
A plant endemic to the west end of the Columbia Gorge, it grows on shady, mossy, rocky banks and hills on both sides of the river.

Synthyris stellata grows to twelve inches tall. Both the stems and the basal leaves are glabrous. The leaves are a deep, shiny green, orbicular in shape, deeply cupped and sharply toothed around the edges. The flower racemes are dense, three to six inches long, and somewhat villous. The four petals are purple, one-quarter of an inch long. The two exserted stamens each have a large blue anther; the pistil is longer than the stamens. It flowers in April and May.

288
Fameflower
(*Talinum spinescens* Torr.).
Purslane Family (*Portulacaceae*).
This succulent plant is endemic to the Columbia Basin, where it grows on basaltic outcrops and scablands in sagebrush deserts. It is known in Oregon from two disjunct sites in Wasco County, and from central Washington.

Talinum spinescens is a perennial herb growing to twelve inches tall. Its green, glabrous, fleshy leaves with their stiff, spiny, persistent midribs, are crowded near the base of the plant. The flowers form in a loose raceme at the top of the glabrous stems. The petals, about one-third of an inch long, are pale red to deep magenta, five in number, and subtended by two sepals. There are twenty to thirty stamens with red filaments and yellow anthers. The ovary is superior. It is in flower from June to August.

289 290

289
Arrow-leaf thelypody, also called the Red purple thelypody because of its rich color
(*Thelypodium eucosmum* Robins.).
Mustard Family (*Brassicaceae*).
This plant is known only from tributaries of the John Day River in Grant and Wheeler counties in Oregon where it grows in dry country, usually near streams or in the shade of juniper trees. It is endangered throughout its range.

Thelypodium eucosmum is perennial, sometimes a biennial which grows to twenty inches in height. It is glabrous and somewhat glaucous throughout. The short-stemmed, oblong basal leaves are generally entire; the leaves on the upper stem are clasping and one to three inches long. The flowers are in an elongated raceme, have four reddish-purple sepals, and four bright lilac-purple petals. The latter are nearly one-half inch long, twice the length of the sepals. The petals, broadest toward the tip, narrow abruptly to a long, slender attachment called a claw. The stamens are slightly longer than the petals. The seed pods tend to curve upward. It blooms May to July.

290
Howell's thelypody
(*Thelypodium howellii* Watson *ssp. spectabilis* [Peck] Al-Shehbaz).
Mustard Family (*Brassicaceae*).
There are historical records of this plant from Baker, Union, Malheur, and Harney counties in eastern Oregon, but it is presently known from only one, privately-owned area near North Powder. It is a candidate for federal listing as endangered. Its habitat has been in river valleys and moist alkaline plains.

Thelypodium howellii ssp. spectabilis grows to over thirty inches tall, and is glabrous and glaucous. The basal leaves are oblanceolate and toothed or lobed. The stem leaves are narrowly lance-shaped with lobes directed backwards clasping the stem. The flowers are in a long, rather loose raceme. Each flower has four sepals and four petals that are purple or lavender in color with wavy, white, thin, membranous margins. The three-quarter inch petals, spatulate in shape, taper to a long, narrow stalk called a claw. The stamens are much shorter than the petals. The flowers bloom in June and July.

291 292

291
Siskiyou Mountains pennycress
(*Thlaspi montanum* L. *var. siskiyouense* P. Holmgren).
Mustard Family (*Brassicaceae*).
This species of moist, rocky serpentine soil is known only in southern Josephine and Curry counties in Oregon. Though currently stable in its range, it needs continued monitoring. Another variety of this species, *californicum* [Watts.] P. Holmgren, in Humboldt County, California, is listed in that state as rare and endangered.

Thlaspi montanum var. siskiyouense grows from five to ten inches tall. Its foliage is generally glabrous, purple in color, and sometimes coated with a whitish powdery-like substance on the surface. The basal leaves are spatulate, with stems longer than the leaf blade; the upper leaves are sessile and clasping. The flowers are white and form a rather dense raceme at the top of the stem. *Var. siskiyouense* blooms from April into June.

292
Bigleaf clover
(*Trifolium howellii* Wats.).
Pea Family (*Fabaceae*).
This rare species grows in moist meadows and woods, and along streams in southwest Oregon from southern Lane County south into California.

Trifolium howellii, a perennial, has stems up to two feet long. The leaf-like stipules at the base of each leaf are green, with entire margins, and ovate in shape. The leaves have three leaflets, each up to two and one-half inches long, which are ovate to elliptical in shape. The flower heads are round, up to an inch broad. The reflexed corolla is white to yellowish in color, and almost one-half inch long. The calyx teeth are triangular, sharp-pointed, less than one-quarter of an inch long, and often reddish tinged. The seed-pods are mostly one-seeded. It blooms in July and August.

Habitat view

293
Owyhee clover
(*Trifolium owyheense* Gilkey).
Pea Family (*Fabaceae*).
This species known only in Malheur County, Oregon, and Owyhee County, Idaho, grows in volcanic ash and diatomaceous talus in places such as Leslie Gulch and Succor Creek. It is rare and threatened throughout its range.

Trifolium owyheense is a perennial herb which grows to eight inches tall. It is glaucous and glabrous except for the flower head. The leaves are made up of three very stiff, thickened, oval leaflets that are notched in the center of the margin, and lightly toothed by the extension of each leaf vein. Each leaf is green with a crescent of lighter green through its center. Each leaf stem has a pair of leaf-like stipules at its base, similar in texture to the leaves. The calyx is greenish-white, very woolly, with long, narrow, pointed sepals. The corolla is one-half to three-quarters inch long, rose-colored, whitish toward the base. Each seed-pod has two to three spotted seeds. It blooms in May and June.

294
Giant purple trillium
(*Trillium kurabayashii* Freeman).
Lily Family (*Liliaceae*).
This rare trillium of southwest Oregon is known only from the Rogue, Pistol, and Chetco river drainages of Curry County and adjacent California. It grows in moist woods and near streams. It is named for a Japanese scientist who spent much time studying the trillium of Oregon. It was at one time considered to be the same species as the white sessile trillium, *Trillium chloropetalum* [Torr.] Howell; it has only recently been described as separate. It has been placed on review lists in both Oregon and California where it has been seen in Del Norte County. More information is needed to determine its rarity status.

Trillium kurabayashii, reaching twelve to twenty inches in height, is one of the largest trilliums in Oregon. At the top of a stout stem, there is a whorl of three large dark green, mottled leaves, up to six inches long and nearly as wide. The sepals, about one-half the length of the petals, are narrowly-elliptic, pointed, and purplish-green. The petals are sessile, from two to nearly four inches long, and about one-third as wide. They are a rich purple color, widest toward the tip, and often twisted. The six large anthers are dark purple. It blooms in March and April.

Close-up of flower

295
Small-flowered sessile trillium
(*Trillium parviflorum* Soukup).
A newly named species resulting from the studies of northwest trillium by Dr. Soukup of Cincinnati, Ohio. This sessile trillium is found only west of the Cascades, north of Marion and Polk counties in Oregon, and north into western Washington. It grows in moist, shady woods. They are seen in few sites in Oregon, and in Washington they are described as being vulnerable and declining in numbers.

The flower of *Trillium parviflorum* is white and does not have a stem. The three ascending petals are very narrow in comparison to their length, about one and three-quarters of an inch long, and three-eighths of an inch wide. They are almost straight-sided instead of widening in the middle. The three styles are purple on the outside, white on the inner stigmatic side. The stamens have a short filament with long, broad anthers that are curved in cross-section. The leaves are large, rounded, green, and somewhat blotched with a darker green color. They grow nine to twelve inches tall. They bloom in April and May.

296
Yellow triteleia
(*Triteleia crocea* [Wood] Greene), until 1987 it was known as *Brodiaea crocea*, (Wood) Wats., named for a Scotch botanist, James Brodie.
Lily Family (*Liliaceae*).
It grows only in the open woods in the Siskiyou Mountains of southern Jackson County in Oregon, south into Del Norte, Shasta, Siskiyou and Trinity counties, California. It is rare throughout its range and needs continued monitoring.

Triteleia crocea has a slender stem ten inches tall, arising from a fibrous-coated corm. Its leaves are basal and shorter than the flower stem. The stem is topped with an umbel of five to fifteen flowers, each about three-quarters of an inch long, in close arrangement. They are bright yellow with prominent deep green to brownish-purple mid-veins. The stamens are in two rows. The filaments of the lower ones are almost non-existent, the upper ones have longer, dilated filaments. The white anthers are about one-eighth of an inch long. It blooms in June and July.

297
Leach's brodiaea
(*Triteleia hendersonii* Greene *var. leachiae* [Peck] Hoover, *Brodiaea hendersonii* Wats. *var. leachiae*).
Lily Family (*Liliaceae*).
Named in honor of its discoverer, Lilla Leach of Portland. This rare species is found only on open or wooded slopes in the Siskiyou Mountains of Josephine, Curry and Douglas counties in Oregon. Though currently stable, it is definitely limited in abundance.

Triteleia hendersonii var. leachiae reaches a height of about ten inches. It has two narrow basal leaves that are slightly longer than the flowering stem. The flowers are in a cluster of six to twelve at the top of the stem. The six tepals are white with prominent mid-veins of dark purple. The stamens are of equal length, and have white or blue anthers. It blooms in April and May.

298
Common triplet lily, Ithuriel's Spear
(*Triteleia laxa* Benth, or *Brodiaea laxa* [Benth.] Wats.).
Lily Family (*Liliaceae*).
This species of the Sierra and central Coast Ranges of California reaches the northern edge of its range in the Siskiyou Mountains of Curry and Jackson counties, Oregon. It is considered to be endangered in Oregon but more common in California.

Triteleia laxa grows up to two feet tall, and has leaves sixteen inches long arising from the base of the plant. Its cluster of six to thirty flowers, each up to one and three-quarters inches long, are violet-purple ranging to nearly white, are funnel-form, and taper to a pointed base. It has two rows of stamens attached at different levels. It blooms from June to August.

299

American globeflower

(*Trollius laxus* Salisb. *ssp. albiflorus* [Gray] Love, Love & Kapoor).
Buttercup Family (*Ranunculaceae*).
This plant, known in Oregon only on a few high peaks in the Wallowa Mountains, is also found in the Olympics, northern Cascades and Wenatchee Mountains in Washington. It grows in very wet, poorly drained areas. It is described as endangered in Oregon but more stable elsewhere.

Trollius laxus var. albiflorus reaches eight to twenty inches in height, has long-stemmed, deeply lobed basal leaves, and a few short-stemmed leaves close to the flower. The large flowers consist of five or more white sepals, about three-quarters of an inch long. There are no petals. There are numerous yellow-anthered stamens and greenish-colored pistils. The many-seeded follicles are tubular, sharp-pointed, and one-third of an inch long. It blooms from June to August.

300

Yellow inside-out flower

(*Vancouveria chrysantha* Greene).
Barberry Family (*Berberidaceae*).
This rare regional endemic is found on serpentine rock and soil only in the Siskiyou Mountains of Josephine and Curry counties in southwest Oregon and Del Norte and Siskiyou counties in northern California.

Vancouveria chrysantha has flowering stems that are up to sixteen inches long. They and the leaf stems are covered with long, red, glandular hairs. The leaflets, each about one and one-half inches long, are orbicular, shallowly three-lobed, without hairs on the upper surface but somewhat fuzzy-villous underneath. The flowers are in a loose raceme of four to fifteen at the end of the stem. They are bright yellow, and one-half to three-quarters of an inch long. The six sepals, and six petals are all sharply reflexed. Each petal has a hood-shaped, nectar-producing appendage near its base. There are also six erect stamens. The flowers are in bloom in May and June.

301
Small-flowered vancouveria
(*Vancouveria planipetala* Calloni).
Barberry Family (*Berberidaceae*).
This rare plant of dry, shaded woods, most often is found on serpentine soil, occasionally in Josephine, Curry, and Coos counties in Oregon, and more abundantly in California.

Growing to twenty inches tall, *Vancouveria planipetala* is characterized by its small white "inside-out" flowers, about one-quarter to one-third of an inch long. The long, arching, thread-like stems are covered with red, glandular hairs. The two sets of sepals, about one-sixth of an inch long, are reflexed, and each of the six petals, also sharply reflexed, have a nectariferous appendage near its base. The leaves are shallowly three-lobed, and are a smooth, glossy, deep green on top and pale green and hairless underneath. It blooms in May and June.

302
Western Canada violet
(*Viola canadensis* L. *var. rugulosa*
[Greene] Hitchc.).
Violet Family (*Violaceae*).
In Oregon this plant is found only in the Imnaha River-Horse Creek area of Wallowa County. It is more common further north in Washington, western Canada, Alaska, and east through the Rockies to Colorado, central United States and the southern Appalachian Mountains. It grows in moist woodlands in loamy soil.

Viola canadensis var. rugulosa arises from stolons along the ground which may be buried. It is pubescent with leaves usually wider than long, heart-shaped at the base, pointed at the tip, and ciliated along the margins. The petals are white, aging purplish, somewhat purplish on the back, yellowish at the base, with the lateral pair bearded. The capsules are ovoid and hairy. It blooms May to July.

303
Golden violet
(*Viola douglasii* Steud.).
Violet Family (*Violaceae*).
This rare species is found in Josephine, Jackson, and Klamath counties in southern Oregon, and as far north as Jefferson County on the eastern side of the Cascades. It prefers open places in gravelly soils and grassy areas.

Viola douglasii has stems up to six inches tall. Its lightly hairy leaf-blades are one and one-half inches long, are on petioles up to three times longer, and are divided into numerous narrow lobes. The flowers are a deep golden yellow, the petals are broad and purple-veined at the base; the two upper petals are brownish-purple on the back. It blooms from March to May.

304
Hall's violet
(*Viola hallii* Gray).
Violet Family (*Violaceae*).
This rarely-seen small violet is found west of the Cascade Mountains from Clackamas and Marion counties to Josephine and Curry counties in Oregon, and south into California. It grows in gravelly soil, grasslands and open woods.

Viola hallii grows two to four inches tall. The two upper petals, three-eighths of an inch long, are deep purple on both sides, the lower three are white to cream color with purple lines and bright yellow bases. The spur is very short. The leaves are hairless, somewhat powdery in appearance and unequally divided into linear or lance-shaped segments on stems two to four inches long. It blooms from March to June.

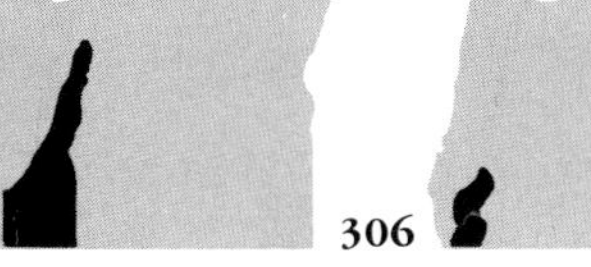

305
Howell's violet
(*Viola howellii* Gray).
Violet Family (*Violaceae*).
Another rarely-seen violet, it prefers moist woods and prairies along the west side of the Cascades, south to California, east to Klamath Lake, and north to British Columbia.

Viola howellii has deep to light blue petals (white at Klamath Lake) that fade to white near the base and are about one-half of an inch long. The lateral petals have tufts of white hair near their base. There is a relatively broad spur about one-third the length of the petals. The leaves are broad, heart-shaped or kidney-shaped, are on long stems, and are sparsely hairy along the veins. The stipules at the base of the leaf stems are sharply toothed and glandular. The plant grows to about four inches tall. It blooms in April and May.

306
Western bog violet
(*Viola lanceolata* L. *ssp. occidentalis* [Gray] Russell, also known as *Viola occidentalis* [Gray] Howell.)
Violet Family (*Violaceae*).
This plant, rare and threatened in Oregon, grows in *Darlingtonia* and sphagnum bogs and swamps in serpentine soil in Curry and Josephine counties, and in adjacent Del Norte County, California where it is considered to be endangered.

Viola lanceolata ssp. occidentalis grows to six inches tall. The leaves are glabrous, broad-lanceolate in shape, and on hairless stems that are longer than the leaf blades. The petals, about one-half of an inch in length, are pure white on both sides. The lower petal has purple veins at the base; the lateral petals are bearded. The spur is short and sac-shaped. It blooms from April to early June.

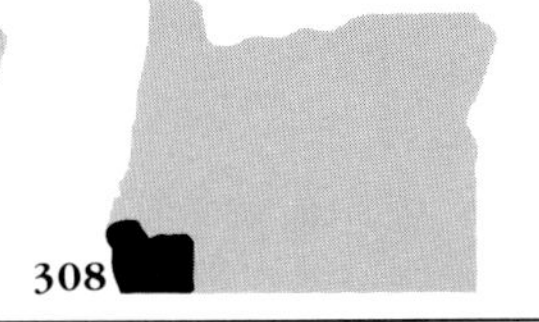

307
Eyed violet
(*Viola ocellata* Torr. & Gray).
Violet Family (*Violaceae*).
This small California violet can be found at scattered sites in the woods of Douglas, Curry, and Josephine counties in Oregon.

Viola ocellata, growing from two to ten inches tall, is a glabrous or very slightly pubescent plant with broadly ovate, pointed leaves, that are from three-quarters to one and one-half inches long, and are toothed on the margins. The basal leaves have long stems; the upper leaves are smaller and have short stems. There are lance-shaped stipules at the base of the leaves. The petals are white, and less than one-half of an inch long. The lower petal is purple-veined, with a blotch of yellow at the base; the lateral have dark purple "eyes" banded with yellow; the upper two are dark purple on the back. The spur is very short and broad. There are small, undeveloped flowers in the upper leaf axils. It blooms from April to June.

308
Broad-leaved California fuchsia
(*Zauschneria latifolia* [Hook.] Greene or *Epilobium canum* [Greene] Raven *ssp. latifolia* [Hook.] Raven).
Evening Primrose Family (*Onagraceae*).
This showy plant of the mountains of California and Nevada can be found as far north as Curry and Josephine counties, Oregon. It grows on dry slopes and ridges and along streams.

Zauschneria latifolia has slender stems that grow as tall as twenty inches. The sessile leaves, up to one and one-half inches long, are numerous, mostly opposite, broadly lance-shaped, sharp-pointed, and have sharply toothed to nearly entire margins. The flowers are bright scarlet with funnel-shaped tubular calyces which are glandular and swollen at the base. The petals are two-cleft and over one-half of an inch long. The stamens are well exserted; the style of the pistil is much longer than the stamens. It blooms from June to September.

APPENDIX

Status of Degree of Rarity by State and Federal Governments

Methods used to determine the degree of rarity of a native plant vary from state to state, and are to various degrees related to the listings of the Federal Endangered Species Act of 1973. Determinations generally are made by botanical organizations or committees established for this purpose. Rarity is based upon historic information plus current reports of findings by botanists. Special field trips for the primary purpose of searching for both new and historic sites also help determine the number of populations and the number of individuals of a particular population. The threats present, which also influence its stability, are assessed.

Oregon's original list of rare, threatened and endangered plants was determined by the Advisory Committee of professional taxonomists of the Oregon Rare and Endangered Plant Project. This list was published by the Natural Area Preserves Advisory Committee to the State Land Board in 1979 as *Rare, Threatened and Endangered Vascular Plants in Oregon—an Interim Report.*

For the past six years Oregon's rare species determination has been managed by the Oregon Natural Heritage Data Base of The Nature Conservancy with the help of many other groups and individuals. Their most recent publication, dated April 1989, is entitled, *Rare, Threatened and Endangered Plants and Animals of Oregon.* In this publication plants have been categorized into six lists in Oregon:

List 1: Plants endangered, threatened, or possibly extinct throughout their entire range. (123 taxa in 1989).

List 2: Plants endangered, threatened, or possibly extinct in Oregon but more common or stable elsewhere. (178)

List 3: Plants that are limited in abundance throughout their range but are currently stable. (63)

Review List: Species for which more information is needed before status can be determined, but which may be endangered in Oregon or throughout their range. (137)

Watch List: Species which are currently stable but which may become threatened in the for foreseeable future. They need some type of continued monitoring. (95)

Drop List: All taxa deleted from any of the above lists after 1979, when the *Rare, Threatened and Endangered Vascular Plants in Oregon—an Interim Report was published.* (168)

With the passage of new endangered species legislation in Oregon in 1987, responsibility for plants in Oregon which are endangered or threatened throughout their range is vested in the Oregon Department of Agriculture.

The state of Washington rare plant program is administered by the Washington Natural Heritage Program under the Washington State Department of Natural Resources. Their most recent listing, entitled *Endangered, Threatened and Sensitive Vascular Plants of Washington* is dated June 1987. Rare plant species are placed under one of the following headings:

Endangered: A vascular plant taxon in danger of becoming extinct or extirpated in Washington within the near future if factors contributing to its decline continue. (10 taxa)

Threatened: A vascular plant taxon likely to become endangered within the near future in Washington if factors contributing to its population decline continue. (41)

Sensitive: A vascular plant taxon that is vulnerable or declining, and could become endangered or threatened in the state. (192)

A monitor list containing species needing more field study, or having name problems, or which are more abundant in Washington than once believed. (138)

The rare plants of Idaho are listed in the June 1981 book, *Vascular Plant Species of Concern in Idaho*, published by the Forest, Wildlife and Range Experiment Station, University of Idaho, Moscow, Idaho. The rare plants are listed in basically two sections or categories:

Section A: Taxa of Federal concern: Basically those Idaho plants (68 taxa) that were listed in the Federal Register plus a few that merit Federal consideration.

Section B: Taxa of State concern: Idaho plants that merit special consideration at the State level, as endangered, threatened, or in need of monitoring. (139)

The Northern Nevada Native Plant Society revised the *Sensitive Plant List for Nevada* in April 1987. The 186 plants listed are placed in one of the following categories:

- Federal Endangered: Nevada plants listed in Federal Register. (2 species)
- Possibly Extinct: Presumed extinct, searching continues (1). Recommended for Federal Endangered: Recommendations by Northern Nevada Native Plant Society (5).
- Federal Threatened: Listed in Federal Register (6). Recommended for Federal Threatened: Recommendations by Northern Nevada Native Plant Society (31).
- Watch List: Plants of uncertain abundance and distribution (76).
- New Taxa: Newly described, little distributional information (1).
- Other Rare: Plants of limited distribution, not now threatened (64).

The California Native Plant Society published in September 1988 an *Inventory of Rare and Endangered Vascular Plants of California*. The listings in the California book are in the following categories:

List 1A: Plants presumed Extinct in California. (39 taxa)

List 1B: Plants Rare and Endangered in California and elsewhere. (675)

List 2: Plants Rare or Endangered in California, but More Common Elsewhere. (177)

List 3: Plants About Which We Need More Information. (149)

List 4: Plants of Limited Distribution (a Watch List). (508)

The Federal Lists resulting from the enactment of the Endangered Species Act of 1973, consist of the following groups. The number of Oregon taxa listed as of April 1987 are given:

Determined to be Endangered: (3 species)

Candidate Species, Category 1: "Sufficient information on hand to support the biological appropriateness of their being listed, but delayed for lack of environmental and economic impact data." (7 species)

Candidate Species, Category 2: "Taxa for which information indicates the probable appropriateness of listing as Endangered or Threatened, but for which sufficient information is not presently available to biologically support a proposed rule. (135)

Rarity Status List for the Species in this Volume

Abbreviations used:

Oregon:

OR 1 EN	List 1, Endangered Throughout Range
OR 1 TH	List 1, Threatened Throughout Range
OR POEX	Possibly Extinct or Extirpated from Oregon
OR 2 EN	List 2, Endangered in Oregon, More Common Elsewhere
OR 2 TH	List 2, Threatened in Oregon, More Common Elsewhere
OR 3	List 3, Limited in Abundance Throughout Range but Currently Stable
OR WL	Watch List
OR RL	Review List
OR DL	Dropped since 1979, when Oregon Interim Report was Published
OR RNL	Reviewed at one time but not listed

Washington:

WA EN	Endangered
WA TH	Threatened
WA SE	Sensitive
WA MO	Monitor List

Idaho:

ID FE	Federal Endangered
ID FT	Federal Threatened
ID FWL	Federal Watch List
ID SE	State Endangered
ID ST	State Threatened
ID SWL	State Watch List

Nevada:

NV FE	Federal Endangered
NV PE	Possibly Extinct
NV RFE	Recommended for Federal Endangered
NV FT	Federal Threatened
NV RFT	Recommended for Federal Threatened
NV WL	Watch List

California:

CA 1A	Plants Presumed Extinct in California
CA 1B	Rare and Endangered in California and Elsewhere
CA 2	Rare or Endangered in California, More Common Elsewhere
CA 3	Plants for Which More Information is Needed
CA 4	Plants of Limited Distribution (Watch List)

Federal:

FE EN	Listed Endangered
FE C1	Category 1 Candidate, Enough Information on Hand to List
FE C2	Category 2 Candidate, More Information/Study Needed

The Plants

Abronia umbellata ssp. breviflora	**OR 1 EN FE C2**
Agoseris elata	**OR 2 EN WA SE**
Alisma gramineum var. angustissimum	**OR RNL**
Allium bolanderi	**OR 2 TH**
Allium douglasii var. nevii	**OR WL**
Allium macrum	**OR WL**
Allium madidum	**OR 3 ID FWL**
Allium platycaule	**OR WL**
Allium pleianthum	**OR 3**
Allium unifolium	**OR WL**
Anemone oregana var. felix	**OR 2 EN WA MO**
Arabis aculeolata	**OR WL CA 2**
Arabis breweri	**OR DL**
Arabis koehleri var. koehleri	**OR 1 EN FE C2**
Arabis koehleri var. stipitata	**OR WL FE C2**
Arabis macdonaldiana	**OR 1 TH CA 1B FE EN**
Arabis modesta	**OR 3 CA 3**
Arenaria Californica	**OR WL**
Arenaria howellii	**OR RL CA 4**
Arenaria pumicola	**OR DL**
Argemone munita ssp. rotundata	**OR 2 TH**
Arnica amplexicaulis var. piperi	**OR DL**
Arnica cernua	**OR DL CA 4**
Artemisia ludoviciana ssp. estesii	**OR 1 EN FE C2**
Asarum wagneri	**OR 3**
Asplenium septentrionale	**OR 2 EN CA 2**
Aster curtus	**OR 2 TH WA SE FE C2**
Aster gormanii	**OR 1 TH FE C2**
Aster vialis	**OR 1 TH FE C2**
Astragalus applegatei	**OR 1 EN FE C2**
Astragalus hoodianus	**OR 3 WA SE**
Astragalus misellus	**OR DL**
Astragalus peckii	**OR 1 TH FE C2**
Astragalus reventus var. reventus	**OR RNL**
Astragalus sterilis	**OR 1 EN ID FE FE C2**
Astragalus tyghensis	**OR 1 EN FE C2**
Balsamorhiza sericea	**OR WL CA 4 FE C2**
Bensoniella oregana	**OR 3 CA 1B FE C2**
Bolandra oregana	**OR 3 WA SE**
Botrychium pumicola	**OR 1 EN CA 1A FE C2**
Calochortus greenei	**OR 1 TH CA 1B FE C2**
Calochortus howellii	**OR 1 TH FE C2**
Calochortus longebarbatus var. peckii	**OR 3 FE C2**
Calochortus bruneaunis	**OR DL**
Calochortus umpquaensis	**OR 1 EN FE C2**
Calypso bulbosa	**OR RNL**
Camassia cusickii	**OR WL ID FWL**
Camassia leichtlinii var. leichtlinii	**OR DL**
Cardamine pattersonii	**OR 1 TH FE C1**
Cardamine penduliflora	**OR DL**
Castilleja chlorotica	**OR 1 TH FE C2**
Castilleja elata	**OR DL CA 2**
Castilleja glandulifera	**OR WL**
Castilleja levisecta	**OR POEX WA EN FE C2**
Castilleja miniata var. dixonii	**OR RL**
Castilleja pilosa var. steenensis	**OR 3 FE C2**
Castilleja rupicola	**OR RL**
Castilleja xanthrotricha	**OR WL FE C2**
Caulanthus crassicaulis	**OR WL**
Chaenactis douglasii var. glandulosa	**OR DL WA SE**
Chaenactis nevii	**OR 3**
Chaenactis stevioides	**OR 2 TH**
Cimicifuga laciniata	**OR DL W MO**
Cirsium acanthodontum	**OR DL**
Cirsium peckii	**OR DL**
Cladothamnus pyrolaeflorus	**OR DL**
Claytonia nevadensis	**OR WL**
Claytonia umbellata	**OR WL**
Clematis columbiana var. columbiana	**OR RL**
Clintonia andrewsiana	**OR 2 EN**
Collomia debilis var. larsenii	**OR WL CA 2**
Collomia macrocalyx	**OR 3 FE C2**
Collomia mazama	**OR 3 FE C2**
Corallorhiza trifida	**OR WL** (Western Oregon)
Cordylanthus maritimus ssp. palustris	**OR 1 TH CA 1B FE C2**
Corydalis aquae-gelidae	**OR 3 WA TH FE C2**
Coryphantha vivipara	**OR RL ID SWL**
Crepis bakeri ssp. cusickii	**OR DL**

Cupressus bakeri ssp. matthewsii	**OR 2 EN CA 3**	*Erigeron howellii*	**OR 3 WA TH FE C2**
Cymopterus bipinnatus	**OR 2 TH**	*Erigeron peregrinus ssp. peregrinus var. peregrinus*	**OR 2 TH**
Cypripedium calceolus var. parviflorum	**OR POEX WA EN ID FE**	*Eriogonum chrysops*	**OR 1 EN FE C2**
Cypripedium californicum	**OR 3 CA 4**	*Eriogonum cusickii*	**OR 1 TH FE C2**
Cypripedium fasciculatum	**OR 2 TH WA TH ID SWL CA 4**	*Eriogonum diclinum*	**OR 2 EN CA 4**
Cypripedium montanum	**OR WL CA 4**	*Eriogonum umbellatum var. hausknechtii*	**OR DL**
Darlingtonia californica	**OR WL CA 4**	*Erythronium citrinum*	**OR DL CA 4**
Delphinium leucophaeum	**OR 1 TH WA TH FE C2**	*Erythronium elegans*	**OR 1 TH FE C2**
Delphinium nuttallii	**OR RL WA MO**	*Erythronium howellii*	**OR 2 TH CA4**
Delphinium pavonaceum	**OR 1 TH FE C2**	*Erythronium klamathense*	**OR DL CA 4**
Dentaria gemmata	**OR 1 TH CA 1B**	*Erythronium oregonum*	**OR RNL WA MO**
Dicentra formosa ssp. oregana	**OR WL CA 4**	*Erythronium revolutum*	**OR WL WA SE**
Dichelostemma ida-maia	**OR WL**	*Eschscholtzia caespitosa*	**OR 2 EN**
Dimeresia howellii	**OR RNL ID ST CA 4**	*Filipendula occidentalis*	**OR 3 WA SE**
Dodecatheon dentatum	**OR DL ID SWL**	*Frasera umpquaensis*	**OR 3 CA 1B FE C2**
Dodecatheon poeticum	**OR DL WA MO**	*Fritillaria adamantina*	**OR DL**
Douglasia laevigata var. ciliolata	**OR RL**	*Fritillaria camschatcensis*	**OR 2 EN WA SE**
Douglasia laevigata var. laevigata	**OR RL WA MO**	*Fritillaria falcata*	**OR RL CA 1B**
Downingia laeta	**OR RL**	*Fritillaria gentneri*	**OR 1 EN FE C2**
Downingia yina var. yina	**OR RL**	*Fritillaria glauca*	**OR 2 Th**
Draba aureola	**OR 3 WA MO CA 1B**	*Fritillaria recurva*	**OR DL**
Draba howellii	**OR 2 TH CA 4**	*Gentiana bisetaea*	**OR 1 TH FE C2**
Draba sphaeroides var. cusickii	**OR 3**	*Gentiana newberryi*	**OR 2 TH**
Dryas drummondii	**OR WL WA SE**	*Gentiana prostrata*	**OR 2 TH CA 4**
Dryas octopetala var. hookeriana	**OR DL**	*Geum triflorum var. campanulatum*	**OR 2 EN WA MO**
Dudleya farinosa	**OR 2 TH**	*Hackelia hispida*	**OR DL WA SE** *(var. disjuncta)* **WA POEX** *(var. hispida)*
Eburophyton austiniae	**OR RNL WA MO ID SWL**	*Haplopappus uniflorus ssp. linearis*	**OR WL**
Elmera racemosa var. puberulenta	**OR WL**	*Haplopappus whitneyi ssp. discoideus*	**OR 2 TH**
Empetrum nigrum	**OR DL**	*Hastingsia atropurpurea*	**OR 1 EN FE C1**
Ephedra nevadensis	**OR DL**	*Hastingsia bracteosa*	**OR 1 EN FE C1**
Ephedra viridis	**OR DL**	*Hedysarum boreale*	**OR RL**
Epilobium latifolium	**OR RL**	*Hemitomes congestum*	**OR DL**
Epilobium oreganum	**OR 1 EN CA 4 FE C2**	*Hieracium bolanderi*	**OR 2 TH**
Epilobium rigidum	**OR 2 TH CA 4**	*Hieracium longiberbe*	**OR 3 WA MO**
Erigeron chrysopsidis var. brevifolius	**OR DL FE C2**	*Horkelia daucifolia*	**OR DL**
Erigeron decumbens ssp. decumbens	**OR 1 EN FE C2**	*Horkelia hendersonii*	**OR 1 TH FE C2**
		Howellia aquatilis	**OR POEX WA EN CA POEX FE C2**
		Hulsea algida	**OR WL**
		Hydrophyllum capitatum var. thompsonii	**OR DL**
		Hymenopappus filifolius var. filifolius	**OR WL**

Hymenoxys cooperi var. canescens	**OR 2 EN**	*Lupinus aridus ssp. ashlandensis*	**OR 1 EN FE C2**
Iliamna latibracteata	**OR 2 TH CA 4**	*Lupinus burkei ssp. caeruleomontanus*	**OR WL**
Iris tenuis	**OR RNL**		
Isoetes howellii	**OR DL**	*Lupinus lyallii ssp. minutifolius*	**OR 3**
Isoetes nuttallii	**OR DL WA SE**		
Isopyrum hallii	**OR DL WA MO**	*Lupinus sabinii*	**OR 3 WA TH**
Ivesia baileyi	**OR RNL ID SWL**	*Lupinus sulphureus var. kincaidii*	**OR 1 TH WA TH**
Ivesia rhypara	**OR 1 EN FE C2**		
Kalmiopsis leachiana var. leachiana	**OR RL**	*Lupinus uncialis*	**OR DL ID SWL**
		Lycopodium sitchense	**OR RNL ID SWL**
Kalmiopsis leachiana var. nov. (Umpqua)	**OR RL**	*Machaerocarpus californicus*	**OR RL ID SWL**
		Malacothrix torreyi	**OR WL**
Lasthenia macrantha ssp. prisca	**OR 3 FE C2**	*Meconella oregana*	**OR 3 WA SE**
		Mentzelia packardiae	**OR 1 EN FE C2**
Lasthenia minor ssp. maritima	**OR DL WA MO**	*Microcala quadrangularis*	**OR 2 TH**
		Microseris howellii	**OR 1 TH FE C2**
Lathyrus delnorticus	**OR 2 TH CA 4**	*Mimulus clivicola*	**OR 1 TH**
Lesquerella douglasii	**OR RL**	*Mimulus douglasii*	**OR 2 TH**
Lesquerella kingii ssp. diversifolia	**OR WL**	*Mimulus jungermannioides*	**OR 3 WA POEX**
		Mimulus kelloggii	**OR 2 TH**
Leucothoe davisiae	**OR DL**	*Mimulus patulus*	**OR 1 TH FE C2**
Lewisia columbiana var. columbiana	**OR 2 TH ID SWL**	*Mimulus pulsiferae*	**OR WL WA SE**
		Mimulus pygmaeus	**OR 1 EN CA 1B**
Lewisia columbiana var. rupicola	**OR 2 TH**	*Mimulus tricolor*	**OR 2 TH**
		Mimulus washingtonensis var. washingtonensis	**OR 1 TH**
Lewisia cotyledon ssp. howellii	**OR WL CA 3 FE C2**	*Mirabilis bigelovii var. retrorsa*	**OR WL**
Lewisia leana	**OR 2 TH**		
Lewisia oppositifolia	**OR 3 CA 1B**	*Mirabilis macfarlanei*	**OR 1 EN ID FE FE EN**
Lilium bolanderi	**OR DL CA 4**		
Lilium kelloggii	**OR POEX**	*Monardella purpurea*	**OR 2 TH CA 4**
Lilium occidentale	**OR 1 EN CA 1B**	*Nama lobbii*	**OR WL**
Lilium parvum	**OR POEX**	*Oenothera wolfii*	**OR 1 TH CA 1B FE C2**
Lilium rubescens	**OR POEX CA 4**		
Lilium vollmeri	**OR DL CA 4**	*Orthocarpus cuspidatus*	**OR WL CA 4**
Lilium wigginsii	**OR WL CA 4**	*Parnassia fimbriata var. hoodiana*	**OR DL WA SE**
Limnanthes floccosa ssp. grandiflora	**OR 1 TH FE C2**		
		Pedicularis bracteosa var. pachyrhiza	**OR DL WA MO**
Limnanthes floccosa ssp. pumila	**OR 1 EN FE C1**		
		Pedicularis densiflora	**OR DL**
Limnanthes gracilis var. gracilis	**OR 3 FE C2**	*Pediocactus simpsonii var. robustior*	**OR WL WA MO ID SWL**
Lloydia serotina	**OR RL WA MO**	*Penstemon barrettiae*	**OR 1 TH WA TH FE C2**
Lomatium bradshawii	**OR 1 EN FE EN**		
Lomatium columbianum	**OR DL**	*Penstemon davidsonii var. praeteritus*	**OR 3**
Lomatium cookii	**OR 1 EN**		
Lomatium greenmanii	**OR 1 EN FE C1**	*Penstemon eriantherus var. argillosus*	**OR WL**
Lomatium howellii	**OR DL CA 4**		
Lomatium laevigatum	**OR 3 WA SE FE C2**	*Penstemon newberryi ssp. berryi*	**OR DL**
Luina serpentina	**OR 1 EN FE C1**	*Penstemon peckii*	**OR 3 FE C2**

Species	Status
Penstemon seorsus	**OR DL**
Penstemon spatulatus	**OR WL FE C2**
Phacelia argentea	**OR 3 CA 1B FE C2**
Phacelia corymbosa	**OR DL**
Phacelia lutea var. mackenzieorum	**OR 1 TH**
Phacelia verna	**OR WL FE C2**
Plagiobothrys hirtus	**OR 1 EN FE C2**
Pleuricospora fimbriolata	**OR DL WA SE**
Polystichum lemmonii	**OR WL**
Potentilla glandulosa ssp. globosa	**OR DL**
Primula cusickiana	**OR 2 TH**
Ranunculus reconditis	**OR 1 EN WA TH FE C2**
Rhamnus crocea ssp. ilicifolia	**OR 2 EN**
Rhinanthus crista-galli	**OR WL**
Ribes cognatum	**OR DL WA SE**
Ribes marshallii	**OR 2 TH CA 4**
Romanzoffia tracyi	**OR DL WA MO**
Sagittaria latifolia	**OR WL**
Salix arctica	**OR WL WA MO**
Salix cascadensis	**OR WL**
Salix drummondiana	**OR WL**
Sanicula peckiana	**OR DL CA 4**
Sarcodes sanguinea	**OR DL**
Saxifraga fragarioides	**OR RL**
Saxifraga hitchcockiana	**OR 1 TH FE C2**
Saxifraga oppositifolia	**OR DL**
Scheuchzeria palustris var. americana	**OR 2 TH ID SWL CA 1B**
Scoliopus hallii	**OR DL**
Sedum laxum ssp. heckneri	**OR 3 CA 4**
Sedum moranii	**OR 1 TH FE C2**
Sedum obtusatum ssp. retusum	**OR DL**
Senecio ertterae	**OR 1 EN FE C1**
Senecio hesperius	**OR 1 TH FE C2**
Sesuvium verrucosum	**OR 2 TH**
Sidalcea campestris	**OR WL FE C2**
Sidalcea cusickii	**OR WL**
Sidalcea hirtipes	**OR 3 WA TH**
Sidalcea malvaeflora ssp. elegans	**OR RL CA 4**
Sidalcea nelsoniana	**OR 1 EN FE C2**
Sidalcea setosa ssp. querceta	**OR POEX**
Sidalcea setosa ssp. setosa	**OR 1 TH CA 4**
Silene douglasii var. oraria	**OR 1 EN FE C2**
Silene hookeri ssp. bolanderi	**OR POEX**
Silene hookeri ssp. pulverulenta	**OR DL CA 3**
Silene suksdorfii	**OR WL**
Sisyrinchium hitchcockii	**OR RL**
Sisyrinchium sarmentosum	**OR 1 EN WA TH FE C1**
Smilax californica	**OR 2 TH**
Sophora leachiana	**OR 3 FE C2**
Spiranthes porrifolia	**OR RL WA SE**
Stephanomeria malheurensis	**OR 1 EN FE EN**
Streptanthus howellii	**OR 3 CA 1B FE C2**
Suksdorfia violacea	**OR 2 EN WA MO**
Sullivantia oregana	**OR 1 EN WA TH FE C2**
Swertia albicaulis var. idahoensis	**OR RNL ID FWL**
Synthyris missurica ssp. hirsuta	**OR WL**
Synthyris schizantha	**OR WL WA MO**
Synthyris stellata	**OR WL WA MO**
Talinum spinescens	**OR 2 TH**
Thelypodium eucosmum	**OR 1 TH FE C2**
Thelypodium howellii ssp. spectabilis	**OR 1 EN FE C1**
Thlaspi montanum var. siskiyouense	**OR WL FE C2**
Trifolium howellii	**OR DL CA 4**
Trifolium owyheense	**OR 1 TH ID FT FE C2**
Trillium kurabayashii	**OR RL CA 3**
Trillium parviflorum	**OR WA SE**
Triteleia crocea	**OR WL CA 4**
Triteleia hendersonii var. leachiae	**OR 3**
Triteleia laxa	**OR 2 EN**
Trollius laxus ssp. albiflorus	**OR 2 EN**
Vancouveria chrysantha	**OR WL CA 4**
Vancouveria planipetala	**OR DL**
Viola canadensis var. rugulosa	**OR DL**
Viola douglasii	**OR DL**
Viola hallii	**OR RNL**
Viola howellii	**OR RNL**
Viola lanceolata ssp. occidentalis	**OR 1 TH CA 1B**
Viola ocellata	**OR DL**
Zauschneria latifolia	**OR DL**

APPENDIX II

Plant Classification of the Species Contained in this Text

Botanical names are used with the most usually accepted common names for each plant family and species given in parenthesis. The order of listing of plant families and genera is based upon the Englerian sequence to show the relationship of the plants to each other pertaining to structure, genetics, and degree of complexity of development. The species, subspecies and variety names are alphabetical.

Phylum I. **Pteridophyta** (Ferns and Fern Allies)

Ophioglossaceae (Adder's-tongue Family)
40. *Botrychium pumicola* (Oregon grape-fern)

POLYPODIACEAE (Fern Family)
238. *Polystichum lemmonii* (Shasta fern)
26. *Asplenium septentrionale* (Grass fern)

ISOETACEAE (Quillwort Family)
158. *Isoetes howellii* (Howell's quillwort)
158. *Isoetes nuttallii* (Nuttall's quillwort)

LYCOPODIACEAE (Club-moss Family)
199. *Lycopodium sitchense* (Alaskan club-moss)

Phylum II. **Spermatophyta** (Seed Plants)

Class I. **GYMNOSPERMAE** (Conifers, Yews, etc.)

CUPRESSACEAE (Cypress Family)
79. *Cupressus bakeri* (Baker's cypress)

EPHEDRACEAE (Ephedra Family)
108. Ephedra nevadensis (Nevada ephedra)
109. Ephedra viridis (Green Mormon tea)

Class II **ANGIOSPERMAE** (Flowering Plants)

Subclass I **MONOCOTYLEDONEAE**

JUNCAGINACEAE (Arrow-grass Family)
256. *Scheuchzeria palustris* (American scheuchzeria)

ALISMACEAE (Water Plantain Family)
3. *Alisma graminea var. angustissimum* (Narrow-leaved water-plantain)
200. Machaerocarpus californicus (Fringed water-plantain)
247. *Sagittaria latifolia* (Wapato)

LILIACEAE (Lily Family)
4. *Allium bolanderi* (Bolander's onion)
5. *Allium douglasii var. nevii* (Nevius' onion)
6. *Allium macrum* (Rock onion)
7. *Allium madidum* (Swamp onion)
8. *Allium platycaule* (Broad-stemmed onion)
9. *Allium pleianthum* (Many-flowered onion)
10. *Allium unifolium* (One-leaved onion)
296. *Triteleia crocea* (Yellow triteleia)
297. *Triteleia hendersonii var. leachiae* (Leach's brodiaea)
298. *Triteleia laxa* (Ithuriel's spear)
91. *Dichelostoma ida-maia* (Firecracker flower)
143. *Hastingsia atropurpurea* (Purple large-flowered rush-lily)
144. *Hastingsia bracteosa* (Large-flowered rush-lily)
47. *Camassia cusickii* (Cusick's camas)
48. *Camassia leichtlinii var. leichtlinii* (White-flowered camas lily)
175. *Lilium bolanderi* (Bolander's lily)
176. *Lilium kelloggii* (Kellogg's lily)
177. *Lilium occidentale* (Western lily)
178. *Lilium parvum* (Alpine lily)
179. *Lilium rubescens* (Lilac lily)
180. *Lilium vollmeri* (Vollmer's lily)
181. *Lilium wigginsii* (Wiggin's lily)
130. *Fritillaria adamantina* (Diamond Lake fritillaria)
131. *Fritillaria camschatcensis* (Kamchatka fritillary)
132. *Fritillaria falcata* (Falcate fritillary)
133. *Fritillaria gentneri* (Gentner's fritillaria)
134. *Fritillaria glauca* (Siskiyou fritillaria)
135. *Fritillaria recurva* (Scarlet fritillary)
121. *Erythronium citrinum* (Lemon-colored fawn-lily)
122. *Erythronium elegans* (Coast Range fawn-lily)
123. *Erythronium howellii* (Howell's adder's-tongue)
124. *Erythronium klamathense* (Klamath fawn-lily)
125. *Erythronium oregonum* (Oregon fawn-lily)
1·26. *Erythronium revolutum* (Coast trout-lily)

185. *Lloydia serotina* (Alp lily)
41. *Calochortus greenei* (Green's mariposa-lily)
42. *Calochortus howellii* (Howell's calochortus)
43. *Calochortus longebarbatus var. peckii* (Long-haired mariposa)
44. *Calochortus bruneaunis* (Sego lily)
45. *Calochortus umpquaensis* (Umpqua mariposa)
257. *Scoliopus hallii* (Oregon fetid adder's-tongue)
294. *Trillium kurabayashii* (Giant purple trillium)
295. *Trillium parviflorum* (Small-flowered trillium)
70. *Clintonia andrewsiana* (Red clintonia)
277. *Smilax californica* (Greenbrier)

IRIDACEAE (Iris Family)
157. *Iris tenuis* (Clackamas iris)
275. *Sisyrinchium hitchcockii* (Hitchcock's purple-eyed grass)
276. *Sisyrinchium sarmentosum* (Pale blue-eyed grass)

ORCHIDACEAE. (Orchid Family)
81. *Cypripedium calceolus var. parviflorum* (Yellow lady's-slipper)
82. *Cypripedium californicum* (California lady's-slipper)
83. *Cypripedium fasciculatum* (Clustered lady's-slipper)
84. *Cypripedium montanum* (Mountain lady's-slipper)
279. *Spiranthes porrifolia* (Western ladies' tresses)
74. *Corallorhiza trifida* (Yellow coral-root)
105. *Eburophyton austiniae* (Phantom orchid)
46. *Calypso bulbosa* (Fairy-slipper)

Subclass II. **DICOTYLEDONEAE**

SALICACEAE (Willow Family)
248. *Salix arctica* (Arctic willow)
249. *Salix cascadensis* (Cascade willow)
250. *Salix drummondiana* (Drummond's willow)

ARISTOLOCHIACEAE (Birthwort Family)
25. *Asarum wagneri* (Green-flowered wild-ginger)

POLYGONACEAE (Knotweed Family)
117. *Eriogonum chrysops* (Golden buckwheat)
118. *Eriogonum cusickii* (Cusick's buckwheat)
119. *Eriogonum diclinum* (James Canyon buckwheat)
120. *Eriogonum umbellatum var. hausknechtii* (Hausknecht's sulfur buckwheat)

NYCTAGINACEAE (Four-o'clock Family)
1. *Abronia umbellata ssp. breviflora* (Pink sand verbena)
215. *Mirabilis bigelovii var. retrorsa* (Desert four-o'clock)
216. *Mirabilis macfarlanei* (Macfarlane's four-o'clock)

AIZOACEAE (Carpet-weed Family)
263. *Sesuvium verrucosum* (Verrucose sea-purslane)

PORTULACACEAE (Purslane Family)
288. *Talinum spinescens* (Fameflower)
67. *Claytonia nevadensis* (Sierra spring beauty)
68. *Claytonia umbellata* (Umbellate spring beauty)
170. *Lewisia columbiana var. columbiana* (Columbia lewisia)
171. *Lewisia columbiana var. rupicola* (Rosy lewisia)
172. *Lewisia cotyledon ssp. howellii* (Howell's lewisia)
173. *Lewisia leana* (Lee's lewisia)
174. *Lewisia oppositifolia* (Opposite-leaved lewisia)

CARYOPHYLLACEAE (Pink Family)
18. *Arenaria californica* (California sandwort)
19. *Arenaria howellii* (Howell's sandwort)
20. *Arenaria pumicola* (Crater Lake sandwort)
271. *Silene douglasii var. oraria* (Cascade Head catchfly)
272. *Silene hookeri ssp. bolanderi* (Bolander's catchfly)
273. *Silene hookeri ssp. pulverulenta* (Hooker's pink)
274. *Silene suksdorfii* (Suksdorf's campion)

RANUNCULACEAE (Buttercup Family)
241. *Ranunculus reconditis* (Obscure buttercup)
11. *Anemone oregana var. felix* (Bog anemone)
69. *Clematis columbiana var. columbiana* (Columbia virgin's-bower)
299. *Trollius laxus var. albiflorus* (American globeflower)
159. *Isopyrum hallii* (Hall's isopyrum)
86. *Delphinium leucophaeum* (White rock larkspur)
87. *Delphinium nuttallii* (Nuttall's larkspur)
88. *Delphinium pavonaceum* (Peacock delphinium)
63. *Cimicifuga laciniata* (Mt. Hood bugbane)

BERBERIDACEAE (Barberry Family)
300. *Vancouveria chrysantha* (Yellow inside-out flower)
301. *Vancouveria planipetala* (Small-flowered vancouveria)

PAPAVERACEAE (Poppy Family)
202. *Meconella oregana* (White meconella)
21. *Argemone munita ssp. rotundata* (Prickly poppy)
127. *Eschscholtzia caespitosa* (Gold poppy)

FUMARIACEAE (Fumitory Family)
90. *Dicentra formosa ssp. oregana* (Oregon dicentra)
76. *Corydalis aquae-gelidae* (Cold-water corydalis)

BRASSICACEAE (Mustard Family)
289. *Thelypodium eucosmum* (Arrowleaf thelypody)
290. *Thelypodium howellii ssp. spectabilis* (Howell's spectacular thelopody)
281. *Streptanthus howellii* (Howell's streptanthus)
59. *Caulanthus crassicaulis* (Thick-stemmed wild cabbage)
291. *Thlaspi montanum var. siskiyouense* (Siskiyou Mountains pennycress)
49. *Cardamine pattersonii* (Saddle Mountain bittercress)

50. *Cardamine penduliflora* (Willamette Valley bittercress)
89. *Dentaria gemmata* (Purple toothwort)
167. *Lesquerella douglasii* (Columbia bladder-pod)
168. *Lesquerella kingii ssp. diversifolia* (King's bladderpod)
99. *Draba aureola* (Golden alpine draba)
100. *Draba howellii* (Howell's whitlow-grass)
101. *Draba sphaeroides var. cusickii* (Cusick's draba)
12. *Arabis aculeolata* (Waldo rock cress)
13. *Arabis breweri* (Brewer's rock cress)
14. *Arabis koehleri var. koehleri* (Koehler's rock cress)
15. *Arabis koehleri var. stipitata* (Koehler's stipitate rock cress)
16. *Arabis macdonaldiana* (Red Mountain rock cress)
17. *Arabis modesto* (Rogue Canyon rock cress)

SARRACENIACEAE (Pitcher-plant Family)

85. *Darlingtonia californica* (California pitcher-plant)

CRASSULACEAE (Stonecrop Family)

258. *Sedum laxum ssp. heckneri* (Heckner's stonecrop)
259. *Sedum moranii* (Glandular stonecrop)
260. *Sedum obtusatum ssp. retusum* (Sierra sedum)
104. *Dudleya farinosa* (Sea-cliff stonecrop)

SAXIFRAGACEAE (Saxifrage Family)

253. *Saxifraga fragarioides* (Strawberry saxifrage)
254. *Saxifraga hitchcockiana* (Saddle Mountain saxifrage)
255. *Saxifraga oppositifolia* (Purple saxifrage)
106. *Elmera racemosa var. puberulenta* (Elmera)
283. *Sullivantia oregana* (Oregon sullivantia)
39. *Bolandra oregana* (Oregon bolandra)
282. *Suksdorfia violacea* (Violet suksdorfia)
38. *Bensoniella oregana* (Bensonia)
221. *Parnassia fimbriata var. hoodiana* (Mt. Hood grass-of-parnassus)

GROSSULARIACEAE (Gooseberry Family)

244. *Ribes cognatum* (Umatilla gooseberry)
245. *Ribes marshallii* (Applegate gooseberry)

ROSACEAE (Rose Family)

239. *Potentilla glandulosa ssp. globosa* (Globose sticky cinquefoil)
149. *Horkelia daucifolia* (Carrot-leaved horkelia)
150. *Horkelia hendersonii* (Henderson's horkelia)
160. *Ivesia baileyi* (Bailey's ivesia)
161. *Ivesia rhypara* (Grimy ivesia)
128. *Filipendula occidentalis* (Queen-of-the-forest)
102. *Dryas drummondii* (Drummond's mountain avens)
103. *Dryas octopetala var. hookeriana* (White mountain avens)
139. *Geum triflorum var. campanulatum* (Western red avens)

FABACEAE (Pea Family)

278. *Sophora leachiana* (Western sophora)
193. *Lupinus aridus ssp. ashlandensis* (Mt. Ashland lupine)
194. *Lupinus burkei ssp. caeruleomontanus* (Blue Mountain lupine)
195. *Lupinus lyallii ssp. minutifolius* (Small-leaved lupine)
196. *Lupinus sabinii* (Sabine's lupine)
197. *Lupinus sulphureus var. kincaidii* (Kincaid's sulfur lupine)
198. *Lupinus uncialis* (Inch-high lupine)
292. *Trifolium howellii* (Bigleaf clover)
293. *Trifolium owyheense* (Owyhee clover)
30. *Astragalus applegatei* (Applegate's milkvetch)
31. *Astragalus hoodianus* (Hood River milkvetch)
32. *Astragalus misellus* (Pauper milkvetch)
33. *Astragalus peckii* (Peck's milkvetch)
34. *Astragalus reventus* (Long-leaf locoweed)
35. *Astragalus sterilis* (Sterile milkvetch)
36. *Astragalus tyghensis* (Tygh Valley milkvetch)
145. *Hedysarum boreale* (Northern sweet-broom)
166. *Lathyrus delnorticus* (Del Norte pea)

EMPETRACEAE (Crowberry Family)

107. *Empetrum nigrum* (Black crowberry)

LIMNANTHACEAE (Meadow-foam Family)

182. *Limnanthes floccosa ssp. grandiflora* (Big-flowered woolly meadow-foam)
183. *Limnanthes floccosa ssp. pumila* (Dwarf meadow-foam)
184. *Limnanthes gracilis var. gracilis* (Slender meadow-foam)

RHAMNACEAE (Buckthorn Family)

242. *Rhamnus crocea ssp. ilicifolia* (Holly-leaved buckthorn)

MALVACEAE (Mallow Family)

264. *Sidalcea campestris* (Meadow sidalcea)
265. *Sidalcea cusickii* (Cusick's sidalcea)
266. *Sidalcea hirtipes* (Bluff mallow)
267. *Sidalcea malvaeflora ssp. elegans* (Checker bloom)
268. *Sidalcea nelsoniana* (Nelson's wild-hollyhock)
269. *Sidalcea setosa ssp. querceta* (Oak Flat sidalcea)
270. *Sidalcea setosa ssp. setosa* (Bristly sidalcea)
156. *Iliamna latibracteata* (California globe-mallow)

VIOLACEAE (Violet Family)

302. *Viola canadensis var. rugulosa* (Western Canada violet)
303. *Viola douglasii* (Golden violet)
304. *Viola hallii* (Hall's violet)
305. *Viola howellii* (Howell's violet)
306. *Viola lanceolata ssp. occidentalis* (Western bog violet)
307. *Viola ocellata* (Eyed violet)

LOASACEAE (Blazing-Star Family)

203. *Mentzelia packardiae* (Packard's mentzelia)

CACTACEAE (Cactus Family)

224. *Pediocactus simpsonii var. robustior* (Hedgehog cactus)
77. *Coryphantha vivipara* (Cushion cactus)

ONAGRACEAE (Evening Primrose Family)

110. *Epilobium latifolium* (Broad-leaved willow-herb)
111. *Epilobium oreganum* (Oregon willow-herb)

112. *Epilobium rigidum* (Rigid willow-herb)
308. *Zauschneria latifolia* (Broad-leaved California fuchsia)
219. *Oenothera wolfii* (Wolf's evening primrose)

APIACEAE (Parsley Family)
251. *Sanicula peckiana* (Peck's snakeroot)
80. *Cymopterus bipinnatus* (Hayden's cymopterus)
186. *Lomatium bradshawii* (Bradshaw's lomatium)
187. *Lomatium columbianum* (Columbia lomatium)
188. *Lomatium cookii* (Agate Desert lomatium)
189. *Lomatium greenmanii* (Greenman's lomatium)
190. *Lomatium howellii* (Howell's lomatium)
191. *Lomatium laevigatum* (Smooth desert parsley)

ERICACEAE (Heath Family)
237. *Pleuricospora fimbriolata* (Fimbriate pinesap)
252. *Sarcodes sanguinea* (Snow plant)
146. *Hemitomes congestum* (Gnome plant)
66. *Cladothamnus pyrolaeflorus* (Copper bush)
162. *Kalmiopsis leachiana* (Kalmiopsis)
163. *Kalmiopsis leachiana var. nov.* (North Umpqua kalmiopsis)
169. *Leucothoe davisiae* (Western leucothoe)

PRIMULACEAE (Primrose Family)
240. *Primula cusickiana* (Wallowa primrose)
95. *Douglasia laevigata var. ciliolata* (Ciliolate douglasia)
96. *Douglasia laevigata var. laevigata* (Smooth-leaved douglasia)
93. *Dodecatheon dentatum* (White shooting star)
94. *Dodecatheon poeticum* (Narcissus shooting star)

GENTIANACEAE (Gentian Family)
204. *Microcala quadrangularis* (Timwort)
136. *Gentiana bisetaea* (Elegant gentian)
137. *Gentiana newberryi* (Newberry's gentian)
138. *Gentiana prostrata* (Moss gentian)
284. *Swertia albicaulis var. idahoensis* (White-stemmed swertia)
129. *Frasera umpquaensis* (Umpqua swertia)

POLEMONIACEAE (Phlox Family)
71. *Collomia debilis* (Larsen's collomia)
72. *Collomia macrocalyx* (Bristle-flowered collomia)
73. *Collomia mazama* (Mt. Mazama collomia)

HYDROPHYLLACEAE (Waterleaf Family)
153. *Hydrophyllum capitatum var. thompsonii* (Ballhead waterleaf)
232. *Phacelia argentea* (Silvery phacelia)
233. *Phacelia corymbosa* (Rock phacelia)
234. *Phacelia lutea var. mackenzieorum* (Mackenzie's yellow phacelia)
235. *Phacelia verna* (Spring phacelia)
218. *Nama lobbii* (Lobb's nama)
246. *Romanzoffia tracyi* (Tracy's mistmaiden)

BORAGINACEAE (Borage Family)
236. *Plagiobothrys hirtus* (Rough allocarya)
140. *Hackelia hispida* (Rough stickseed)

LAMIACEAE (Mint Family)
217. *Monardella purpurea* (Siskiyou monardella)

SCROPHULARIACEAE (Figwort Family)
225. *Penstemon barrettiae (Barrett's penstemon)*
226. *Penstemon davidsonii var. praeteritus (Davidson's penstemon)*
227. *Penstemon eriantherus var. argillosus (Crested tongue penstemon)*
228. *Penstemon newberryi (Mountain pride)*
229. *Penstemon peckii (Peck's penstemon)*
230. *Penstemon seorsus (Short-lobed penstemon)*
231. *Penstemon spatulatus (Wallowa penstemon)*
206. *Mimulus clivicola (Hill monkeyflower)*
207. *Mimulus douglasii (Douglas' monkeyflower)*
208. *Mimulus jungermannioides (Hepatic monkey-flower)*
209. *Mimulus kelloggii (Kellogg's monkeyflower)*
210. *Mimulus patulus (Stalk-leaved monkeyflower)*
211. *Mimulus pulsiferae (Pulsifer's monkeyflower)*
212. *Mimulus pygmaeus (Pygmy monkeyflower)*
213. *Mimulus tricolor (Tri-colored monkeyflower)*
214. *Mimulus washingtonensis (Washington monkeyflower)*
285. *Synthyris missurica ssp. hirsuta (Mountain kittentails)*
286. *Synthyris schizantha (Fringed synthyris)*
287. *Synthyris stellata (Columbia kittentails)*
51. *Castilleja chlorotica (Green-tinged paintbrush)*
52. *Castilleja elata (Slender indian paintbrush)*
53. *Castilleja glandulifera (Sticky indian paintbrush)*
54. *Castilleja levisecta (Golden paintbrush)*
55. *Castilleja miniata var. dixonii (Dixon's paintbrush)*
56. *Castilleja pilosa var. steenensis (Steens Mountain paintbrush)*
57. *Castilleja rupicola (Cliff paintbrush)*
58. *Castilleja xanthotricha (Yellow-haired paintbrush)*
220. *Orthocarpus cuspidatus (Broad-scaled orthocarpus)*
75. *Cordylanthus maritimus ssp. palustris (Saltmarsh birdsbeak)*
222. *Pedicularis bracteosa var. pachyrhiza (Bracted lousewort)*
223. *Pedicularis densiflora (Indian warrior)*
243. *Rhinanthus crista-galli (Yellow rattle)*

CAMPANULACEAE (Harebell Family)
97. *Downingia laeta* (White downingia)
98. *Downingia yina var. yina* (Cascade downingia)
151. *Howellia aquatilis* (Howellia)

ASTERACEAE (Composite Family)
141. *Haplopappus uniflorus* (One-flowered goldenweed)
142. *Haplopappus whitneyi ssp. discoideus* (Whitney's haplopappus)
27. *Aster curtus* (White-topped aster)
28. *Aster gormanii* (Gorman's aster)
29. *Aster vialis* (Wayside aster)
113. *Erigeron chrysopsidis var. brevifolius* (Dwarf golden daisy)

114. *Erigeron decumbens ssp. decumbens* (Willamette Valley daisy)
115. *Erigeron howellii* (Howell's erigeron)
116. *Erigeron peregrinus ssp. peregrinus var. peregrinus* (Wandering daisy)
92. *Dimeresia howellii* (Dimeresia)
37. *Balsamorhiza sericea* (Silky balsamroot)
154. *Hymenopappus filifolius var. filifolius* (Columbia cut-leaf)
164. *Lasthenia macrantha ssp. prisca* (Large-flowered goldfields)
165. *Lasthenia minor ssp. maritima* (Hairy lasthenia)
60. *Chaenactis douglasii var. glandulosa* (Hoary chaenactis)
61. *Chaenactis nevii* (John Day chaenactis)
62. *Chaenactis stevioides* (Broad-flowered chaenactis)
152. *Hulsea algida* (Alpine hulsea)
155. *Hymenoxys cooperi var. canescens* (Cooper's goldflower)
24. *Artemisia ludoviciana ssp. estesii* (Estes' artemisia)
192. *Luina serpentina* (Colonial luina)
22. *Arnica amplexicaulis var. piperi* (Clasping arnica)
32. *Arnica cernua* (Nodding arnica)
261. *Senecio ertterae* (Ertter's senecio)
262. *Senecio hesperius* (Siskiyou butterweed)
64. *Cirsium acanthodontum* (Nelson's thistle)
65. *Cirsium peckii* (Steens Mountain thistle)
205. *Microseris howellii* (Howell's microseris)
280. *Stephanomeria malheurensis* (Malheur wire lettuce)
201. *Malacothrix torreyi* (Torrey's malacothrix)
2. *Agoseris elata* (Tall agoseris)
147. *Hieracium bolanderi* (Bolander's hawkweed)
148. *Hieracium longiberbe* (Long-bearded hawkweed)
78. *Crepis bakeri ssp. cusickii* (Baker's hawksbeard)

Glossary

Achene (Akene) Small, dry, one-seeded, indehiscent fruit.
Acuminate Tapering to a pointed apex.
Acute Tapering to the apex, less tapering than acuminate.
Aerial As roots borne above the ground or water, in the air.
Alternate Leaves born singly and not opposite. One at a node.
Annual A plant completing its life cycle in one year.
Anther The pollen bearing part of a stamen.
Apex The point farthest from the place of attachment.
Apical Pertaining to the apex.
Appendage An attached secondary part to a main structure.
Appressed Lying flat and close to something. Usually refers to hairs.
Aquatic Living in water.
Arcuate Curved like a bow.
Ascending Growing diagonally upward, usually at about 45 degrees.
Auriculate Having ear-shaped appendages.
Axile In the axil, the angle between a plant part and its axis.

Banner The upper, often largest petal of a leguminous type flower.
Basal Forming at the base of a plant.
Beak A hard or firm projection at or near the tip of a plant structure.
Bearded Furnished with long or stiff hairs.
Berry A fleshy, pulpy fruit with immersed seeds.
Bicolored Of two contrasting colors.
Biennial A plant completing its life cycle in two growing seasons.
Bifurcate Divided into two branches.
Bilabiate Two-lipped corolla or calyx.
Bilobate Two-lobed.
Bipinnate Pinnately compound with the primary leaflets again pinnately divided.
Blade The expanded, flattened portion of a leaf or petal.
Bloom A fine waxy, powdery deposit on the surface of a plant.
Boreale Pertaining to the north.
Bract A modified leaf located at the base of a flower or inflorescence.
Bracteole Same as bractlet.
Bractlet A very small bract, sometimes secondary to a larger bract.
Bulb An underground leaf-bud with fleshy scales like an onion.
Bulblet A small bulb, often borne above the ground.

Calyx The combined outer segments, or sepals, of a flower.
Campanulate Bell-shaped.
Canescent With white or gray short hairs.
Capitate In the form or shape of a head, or in a dense, head-like cluster.
Capsule A dry dehiscent fruit made up of more than one carpel.
Carpel A simple pistil.
Catkin A spike-like, often pendulous inflorescence of unisexual flowers.
Cauline Pertaining to the stem.
Channeled Having upturned edges, trough-like.
Chlorophyll The green pigment associated with photosynthesis.
Ciliate With hairs on the margins of leaves or other plant parts.
Ciliolate Ciliate with minute hairs.
Circumboreale Around the northern hemisphere.
Clasping A sessile leaf with lower edges partially surrounding the stem.

Clavate Club-shaped, widest near the apex.
Claw A narrow stalk-like base of a petal.
Cleft Divided about half-way to the midrib or base.
Cleistogamous Self-fertilizing, due to various arrangements that prevent cross-pollination.
Column A group of united filaments as in Malvaceae, or united filaments and style in Orchidaceae.
Coma A tuft of hairs, especially on a seed. Usually long and soft.
Composite Referring to a plant with two different types of flowers in the flower head, or to members of the plant family, *Compositae (Asteraceae).*
Cone Seed-bearing structure of most Gymnosperms.
Confluent Running together, blending into one.
Cordate Heart-shape, the point apical.
Corm A short, thick underground stem with roots below, and leaves or shoots above.
Corolla A collective name for the petals of a flower.
Corrugated Marked with ridges.
Cotyledons The first leaves developed in the seed.

Decumbent A stem reclining on the ground a ways before ascending.
Dehiscent Opening by pores or slits to discharge seeds or contents.
Dentate Having the margin broken by more or less irregular projecting points.
Denticulate Finely dentate.
Diatomaceous Consisting of or containing the fossil remains of microscopic, unicellular, marine or fresh-water algae, having siliceous cell walls.
Dichotomous Forking into two equal branches.
Dicotyledonous Having two cotyledons.
Dilated Broadened out.
Dioecious Flowers unisexual, the staminate and pistillate borne on separate plants.
Diploid Having double the basic number of chromosomes.
Disarticulating The parts separating at maturity.
Discoid Referring to disk flowers. No ray flowers present.
Disjunct Widely separated from like populations.
Disk flowers The central tubular flowers on a composite flower head.
Distal The farthest point from the attachment.
Divergent Growing or spreading in different directions.
Dorsal The back or part farthest away from the supporting axis.

Elliptical A form two to three times as long as wide, with uniform curvature of the sides.
Endemic Confined to a limited geographical area.
Entire Margins without teeth or lobes.
Exserted Projecting beyond a surrounding organ.

Falcate Curved like the blade of a sickle.
Fertile Capable of producing fruit and seeds.
Fibrous Made up of or containing fibers.
Filament The portion of the stamen that supports the anther, or any thread-like body.
Filiform Threadlike, long and slender.
Fimbriate Fringed.
Flexuose Bent at the nodes in a zigzag manner.
Follicle A dry fruit with one carpel, and splitting on one side only.
Forma alba A plant with a marked deficiency in pigmentation.
Frond The leaf of a fern.
Funnelform With the tube widening gradually upward into the limb, like a funnel.

Gabbro A granular igneous rock consisting of labradorite and augite.
Galea The narrow upper lip of the corolla in some genera of Scrophulariaceae.
Gamopetalous The corolla in one piece, not of separate petals.
Gamosepalous The calyx in one piece, not of separate sepals.
Glabrous Smooth, with no hairs.
Gland A secreting surface or structure, or an appendage with the appearance of such an organ.
Glandular Having organs or hairs that secrete a sticky or resinous substance.
Glaucous Surface covered with a whitish waxy substance that readily rubs off.
Globose Spherical.
Glochidiate Barbed at the tip.

Herb A plant without woody stems above ground. Also a plant used for seasoning.
Herbaceous With the charactistics of an herb. Also, leaf-like in texture and color.
Herbage The herbaceous parts of a plant above ground, mainly stems and leaves.

Hirsute With spreading, coarse, but not bristly hairs.
Hood An over-arching structure such as the upper lip of a corolla.
Hypanthium An elongation or enlargement of the floral axis below the calyx, causing the stamens and petals to appear to be borne on the edge of the calyx tube.

Imbricate Overlapping like shingles on a roof.
Indehiscent Not opening by definite lines or pores, persistently closed.
Indusium Scale-like covering of the young spores in ferns.
Inflated Distended or bladder-like.
Inflorescence The flowering part of a plant, often referring to a flower cluster.
Inserted Borne upon.
Involucel A small secondary involucre in the *Umbelliferae (Apiaceae)*.
Involucre A whorl of modified leaves or bracts, united or separate, sub-tending a flower or inflorescence.

Keel A prominent longitudinal ridge.

Laciniate With a series of somewhat uniform lobes on either side of the midrib.
Lanate Woolly, with long, soft interwoven hair.
Lanceolate Lance-shaped. Broadest toward the base, tapering to the apex.
Lateral Borne on the sides of a structure.
Leaflet A secondary blade of a compound leaf.
Legume A dehiscent fruit of *Leguminosae (Fabaceae).*
Ligule Flattened strap-like ray-flowers of Composite family plants.
Limb The expanded upper part of the calyx or corolla of a gamopetalous or gamosepalous plant.
Linear Narrow and flat, with parallel sides like a grass-blade.
Lip The upper or lower division of a bilabiate corolla. Also the upper petal of an orchid.
Lobe A rounded segment of a plant organ such as a leaf.

Metasedimentary A metamorphic rock of sedimentary origin.
Mid-rib The central vein or rib of a petal or leaf.
Monocotyledonous Having but one cotyledon.
Monoecious Flowers unisexual but the staminate and pistillate ones borne on the same plant.
Mycotrophic Plants that do not produce chlorophyll, but depend on the fungi in the soil for food.

Nectariferous Bearing nectar.
Node A swollen or knob-like structure on a stem from which leaves or branches normally originate.

Oblanceolate Inversely lanceolate, broader toward the apex.
Oblique With unsymmetrial sides.
Oblong A leaf form in which the length is considerably greater than the width and the lateral margins are parallel for most of their length.
Obovate Egg-shaped in outline, widest toward the tip.
Obtuse Blunt or rounded at the apex.
Ochroleucous Whitish-yellow or cream-colored.
Opposite Two leaves at a node situated across the stem from each other.
Orbicular A two dimensional figure circular in outline.
Oval Broadly elliptical.
Ovary The enlarged basal part of the pistil that eventually contains the seeds.
Ovate Egg-shaped in outline, widest toward the attachment.
Ovoid Egg-shaped in the exact sense (three-dimensional).
Ovulate Containing the structure that developes into the seed.

Palate A rounded projection on the lower lip of a bilabiate corolla, closing the throat.
Palmate With veins or segments radiating from the same point.
Panicle A flower-cluster in which the primary axis branches into secondary. These may bear pedicels, or may branch again.
Papilla A minute nipple-shaped projection. (pl. papillae).
Pappus Hair-like or feather-like structures or scales representing the calyx in most composite plants.
Parallel-veined Veins of a leaf running parallel to each other to the apex and obscurely connected between.
Parasitic Deriving nourishment from another living organism.
Pedicel The stalk to a single flower of an inflorescence.
Pendent Pendulous.

Pendulous Hanging or drooping. Supported from above.
Perennial A plant living for several years.
Perfect A flower having both functional pistils and functional stamens.
Perianth The calyx and corolla considered together, or either of them if the other is absent.
Peridotite Any of a group of igneous rocks of granitic texture, composed chiefly of olivine with an admixture of various other minerals, but nearly or wholly free from feldspar.
Peripheral Toward the outside, away from center.
Persist Remain attached after like parts would ordinarily fall off.
Petal An individual part of the corolla.
Petiole The stalk of a leaf.
Pinnae Primary divisions of a pinnately compound leaf. Often referring to ferns.
Pinnate Compound leaf supporting leaflets on both sides of an elongated axis.
Pistil The seed producing organ of a plant, consisting of a stigma, style, and ovary.
Pistillate Provided with pistils, stamens lacking.
Pod A dry, dehiscent fruit. Legume.
Pre-Cretaceous Prior to the Cretaceous time period of the Mesozoic era. (About 150 million years ago).
Prickle A rigid, spine-like outgrowth of the outer layer of the bark or covering of a stem.
Prostrate Lying flat on the ground.
Puberulent Covered with fine, short hairs.
Pubescent Covered with hairs, usually short, soft hairs.

Raceme An inflorescence of pedicelled flowers on an elongated stem or axis.
Radiate Spreading from a common center. Referring to ray flowers in composite plants.
Ray flower A strap-shaped or ligulate flower, usually around the outer margin of a Compositae.
Receptacle The somewhat expanded portion of the flower stalk that bears the organs of a flower, or the collected flowers of a head as in Compositae.
Recurved Curved outward or backward.
Reflexed Sharply bent back.
Retrorse Directed backwards toward the base.
Revolute Rolled backward from each margin upon the lower side.
Rhizome An underground elongated stem that serves as a food-storage and as a reproductive organ.
Rib A prominent longitudinal ridge.
Rootstock See Rhizome.
Rosette A dense basal cluster of leaves arranged in a circular manner.
Rotate A wheel-shaped corolla with short tube and wide horizontally flaring limb.

Saccate Sac-like, or pouch-shaped.
Sagittate Arrowhead-shaped.
Saprophytic Living on decaying organic matter.
Scabrous Roughened, or harsh to the touch. Usually caused by stiff hairs or sharp projections.
Scale A thin scarious covering on a plant organ.
Scape A flowering stem without leaves.
Scarious Membranous, thin, dry, somewhat transluscent, not green.
Scorpioid Coiled at the apex, usually of an inflorescence, like the tail of a scorpion.
Segment One of a series of divisions of a corolla, calyx, or leaf-blade.
Sepal One part of the calyx or outer whorl of the floral envelope.
Serrate With sharp teeth directed forward.
Serpentine A common mineral, hydrous magnesium silicate, usually greenish in color.
Sessile Without a stem or stalk.
Seta A bristle-like hair. (pl. setae).
Sheath Usually the basal part of a leaf that folds about or encloses the stem.
Shrub A woody perennial plant, smaller than a tree, often with many basal stems.
Siliques The fruit of a mustard family plant, several times longer than wide.
Spathe A large bract sheathing or enclosing an inflorescence.
Spatulate Broad and rounded at the apex, tapering toward the base.
Spine A rigid, deep-seated, sharp-pointed structure, often a modified branch or leaf.
Sporangia The spore-bearing cases in ferns and fern-related plants.
Spore The small reproductive body in ferns and fern-related plants.
Spur A narrow sac-like prolongation of the base of a calyx or corolla.
Stamen A pollen-bearing organ of a flower, consisting of a filament and anther.
Staminate Bearing stamens only.
Staminodia Imperfectly developed stamen, not bearing pollen.
Stellate Star-shaped. Segments or hairs radiating from a common center.
Sterile Infertile and unproductive.

Stigma The part of the pistil that receives the pollen, usually at or near the apex.
Stipe The stalk-like support of a plant structure. Also the petiole of a fern frond.
Stipitate Having a slender stalk-like base.
Stipules Appendages at each side of the base of a petiole or leaf.
Stock The trunk or main stem of a tree or other plant, as distinguished from roots and branches. It is also a rhizome or rootstock.
Stolon A basal branch that takes root and propagates.
Strigose With short, stiff, appressed hairs.
Strobilus An inflorescence of imbricated scales or bracts such as a pine cone. (Pl. strobili)
Style The stalk-like portion of the pistil, connecting the stigma and ovary.
Subglobose Nearly spherical.
Subtending Situated closely beneath something, often enclosing or embracing it.
Succulent Fleshy, pulpy, juicy.
Sulcate Deeply and longitudinally grooved or furrowed.
Sympatric Capable of occupying the same range without losing identity from cross-pollination or interbreeding.

Talus The sloping mass of rocky fragments at the base of a cliff.
Taproot A single main root bearing secondary roots.
Taxa Plural of taxon.
Taxon A classified entity or group in a formal system of nomenclature.
Taxonomy The systematic distinguishing, ordering, and naming of group types within a subject field.
Tendril A slender modified leaf or stem that coils around or clings to a support.
Tepals Petals and sepals that appear similar and are alternate with each other, often in a plant of the lily family.
Terminal Growing at the end of a branch or stem.
Ternate Arranged in threes.
Throat The upper part of the tube of a corolla or calyx, usually dilated.
Tomentose Dense wool-like covering of matted or tangled hairs of medium length.
Toothed With points and indentations along the margin of a leaf.
Trifoliate A compound leaf with three leaflets.
Triternate Each leaf divided into three segments, each segment again divided into three.
Tube The narrow lower part of a corolla or calyx.
Tuber A short, fleshy, underground portion of a stem such as a potato.
Tubercle Small tuber-like structure.
Tuff (Geol.) A fragmental rock consisting of the smaller kinds of volcanic detritus, usually somewhat stratified.
Tufted Stems in a very close cluster, or having a cluster of hairs or slender outgrowths.

Umbo Small green or dark spots near the tips of the lateral sepals of certain delphiniums.
Umbel A convex or flat-topped inflorescence of flowers all arising from one point.
Undulate With a gently wavy margin.
Unisexual Having only stamens, or only carpels.

Velum A thin membrane covering the sporangium in species of Isoetes.
Vernal Referring to pools existing in the spring that usually dry up later in the season.
Verticil A whorl.
Viable Capable of living.
Villous With long, soft, somewhat wavy hairs.
Viscid Glutinous, sticky or gummy to the touch.

Whorl A verticil. A group of three or more leaves or flowers distributed about the axis at a single node.
Wing A thin, membranous expansion surrounding or bordering a plant structure. Also, a lateral petal of a leguminous flower.
Winged Provided with wings.
Woolly Lanate, with long, soft interwoven hair.

Bibliography

Abrams, L. *An Illustrated Flora of the Pacific States, Washington, Oregon, and California.* 4 vols. Palo Alto: Stanford University Press, 1940-1960.

Becking, Rudolf W. *Pocket Flora of the Redwood Forest.* Covelo, California: Island Press, 1982.

Clark, L. G. *Wild Flowers of the Pacific Northwest from Alaska to Northern California.* Sidney, B.C., Canada: Gray's Publishing Limited, 1976.

Cronquist, Arthur, Arthur H. Holmgren, Noel H. Holmgren, James L. Reveal, Patricia K. Holmgren. *Intermountain Flora, Vascular Plants of the Intermountain West, U.S.A.* Vols. One, Four, and Six. New York: Columbia University Press, 1977-1986.

Department of the Interior, U.S. Fish and Wildlife Service. *Endangered and Threatened Wildlife and Plants.* Washington, D.C.: Department of the Interior Fish and Wildlife Service, 1983.

Elmore, Francis H. *Shrubs and Trees of the Southwest Uplands.* Globe, Arizona: Southwest Parks and Monuments Association, 1981.

Forest Service: USDA Pacific Northwest Region, Siskiyou National Forest. *A Guide: to sensitive plants of the Siskiyou National Forest.* Grants Pass, Oregon: United States Department of Agriculture, Forest Service, 1985.

Forest Service: USDA Pacific Northwest Region, Siskiyou National Forest. *Kalmiopsis Wilderness/Wild Rogue Wilderness,* Map brochure. U.S. Government Printing Office: 1980.

Gilkey, Helen M., and La Rea J. Dennis. *Handbook of Northwestern Plants.* Corvallis, Oregon: Oregon State University Bookstores, Inc., 1980.

Harrington, H. D. and L. W. Durrell. *How to Identify Plants.* Chicago: The Swallow Press Inc., 1957.

Hitchcock, C. L., and A. Cronquist. *Flora of the Pacific Northwest.* Seattle: University of Washington Press, 1974.

Horn, Elizabeth L. *Wildflowers 1. The Cascades.* Beaverton, Oregon: The Touchstone Press, 1972.

Horn, Elizabeth L. *Wildflowers. The Pacific Coast.* Beaverton, Oregon: Beautiful America Publishing Company, 1980.

Jepson, Willis Linn. *A Manual of the Flowering Plants of California.* Berkeley, Los Angeles: University of California Press, 1925.

Jolley, R., and L. Kemp. *Survey of Wildflowers and Flowering Shrubs of the Columbia Gorge.* Portland: Native Plant Society of Oregon, 1984.

Jolley, Russ. *Wildflowers of the Columbia Gorge.* Portland: Oregon Historical Society Press, 1988.

Johnson, J. M. *Handbook of Uncommon Plants in the Salem BLM District.* Salem, Oregon: Salem BLM District, 1980.

Kartesz, John T. and Rosemarie. *A Synonymized Checklist of the Vascular Flora of the United States, Canada, and Greenland,* Volume II *The Biota of North America.* Chapel Hill, NC, The University of North Carolina Press, 1980.

Kinucan, Edith S. and Penney R. Brons. *Wild Wildflowers of the West.* Ketchum, Idaho: Kinucan & Brons, 1985.

Kozloff, Eugene N. *Plants and Animals of the Pacific Northwest.* Seattle: University of Washington Press, 1976.

Larrison, Earl J., Grace W. Patrick, William H. Baker, and James A. Yaich. *Washington Wildflowers.* Portland: Durham & Downey, Inc. for The Seattle Audubon Society. 1977.

Mason, Georgia. *Guide to the Plants of the Wallowa Mountains of Northeastern Oregon.* 2nd Edition. Eugene, Oregon: University of Oregon Press, 1980.

Meinke, R. J. *Threatened and Endangered Vascular Plants of Oregon: An Illustrated Guide.* Portland: U.S. Fish and Wildlife Service, 1982.

Munz, Philip A. and David D. Keck. *A California Flora and Supplement.* Berkeley and Los Angeles: University of California Press, 1959-1968.

Niehaus, T. F. and Ripper, Charles L. *A Field Guide to Pacific States Wildflowers.* Boston: Houghton Mifflin Company, 1976.

Northern Nevada Native Plant Society. *1987 Sensitive Plant Workshop Summary.* Carson City, Nevada: Nevada Natural Heritage Program, 1987.

Oregon Natural Heritage Data Base. *Rare, Threatened and Endangered Plants and Animals of Oregon.* Portland: The Nature Conservancy, 1983.

Oregon Natural Heritage Data Base. *Rare, Threatened and Endangered Plants and Animals of Oregon.* Portland: The Nature Conservancy, 1985.

Oregon Natural Heritage Data Base. *Rare, Threatened and Endangered Plants and Animals of Oregon.* Portland: The Nature Conservancy, 1987.

Oregon Natural Heritage Data Base. *Rare, Threatened and Endangered Plants and Animals of Oregon.* Portland: The Nature Conservancy, 1989.

Peck, M. E. *A Manual of the Higher Plants of Oregon.* 2nd ed. Portland: Binford & Mort, 1961.

Rickett, H. W. *The Northwestern States.* Vol. 5 of *Wildflowers of the United States.* New York: McGraw-Hill, 1971.

Rickett, H. W. *The Central Mountains and Plains.* Vol. 6 of *Wild Flowers of the United States.* New York: McGraw-Hill, 1973.

Siddall, J. L., K. L. Chambers, and D. H. Wagner. *Rare, Threatened and Endangered Vascular Plants in Oregon—an Interim Report.* Salem, Oregon: Oregon State Land Board, 1979.

Smith, James Payne Jr. and Ken Berg. *Inventory of Rare and Endangered Vascular Plants of California, Fourth Edition.* Sacramento, California: California Native Plant Society, 1988.

Steele, Robert, Fred Johnson, and Steve Brunsfeld. *Vascular Plant Species of Concern in Idaho.* Moscow, Idaho: University of Idaho, Forest, Wildlife and Range Experiment Station, 1981.

Underhill, J. E. *Wild Berries of the Pacific Northwest.* Seattle: Superior Publishing Company, 1974.

Washington Natural Heritage Program. *Endangered, Threatened & Sensitive Vascular Plants of Washington.* Olympia, Washington: Washington Natural Heritage Program, 1987.

INDEX

Common and Botanical Names

VENGARICK
9731 SE LINWOOD AVE
MILWAUKIE, OR 97222
503-775-6563